AF541549

Textbook on Agricultural Microbiology

NIPA® GENX ELECTRONIC RESOURCES & SOLUTIONS P. LTD.
New Delhi-110 034

Textbook on Agricultural Microbiology

Bholanath Saha, Ph D
Assistant Professor cum Jr. Scientist
Department of Soil Science
Dr. Kalam Agricultural College, Kishanganj
Bihar Agricultural University, Bihar - 855 107, India

Sushanta Saha, Ph D
Assistant Professor
Department of Agricultural Chemistry and Soil Science
Bidhan Chandra Krishi Viswavidyalaya
P.O. Mohanpur, Dist, Nadia, West Bengal - 741 252, India

Partha Deb Roy, Ph D
Scientist
ICAR-Indian Institute of Water Management
Bhubaneswar, Odisha -751 023, India

Rajiv Rakshit, Ph D
Assistant Professor cum Jr. Scientist
Bihar Agricultural College
Bihar Agricultural University
Sabour, Bhagalpur, Bihar – 813 210, India

Nirmalendu Basak, Ph D
Senior Scientist
Division of Soil & Crop Management
ICAR—Central Soil Salinity Research Institute
Karnal, Haryana – 132 001, India

Md. Shamim, Ph D
Assistant Professor cum Jr. Scientist
Department of Molecular Biology and Genetic Engineering
Dr. Kalam Agricultural College, Kishanganj
Bihar Agricultural University, Bihar - 855 107, India

NIPA® GENX ELECTRONIC RESOURCES & SOLUTIONS P. LTD.
New Delhi-110 034

NIPA® GENX ELECTRONIC RESOURCES & SOLUTIONS P. LTD.

101,103, Vikas Surya Plaza, CU Block
L.S.C.Market, Pitam Pura, New Delhi-110 034
Ph. +91 11 27341616, 27341717, 27341718
E-mail: newindiapublishingagency@gmail.com
Website: www.nipabooks.com

For customer assistance, please contact
Phone: + 91-11-27 34 17 17
Fax: + 91-11-27 34 16 16
E-Mail: feedbacks@nipabooks.com

ISBN: 978-81-19002-48-1

Composed and Designed by NIPA®.

Preface

This textbook is a undergraduate standard Agricultural Microbiology for agricultural students. Microbes may be the most significant life form sharing this planet with plants and animals because of their pervasive presence and their utilization of any available food sources and in other important forms. Beneficial roles of microbes include recycling of organic matter through microbe-induced decay and through digestion and nutrition for the soils helps to the plants. The definition of microbiology, or the study of microorganisms, is frequently given as the study of organisms and agents that are too small to be seen clearly with the unaided eye. Microbiology is primarily concerned with organisms and agents this size and smaller since items less than around one millimetre in diameter cannot be seen properly and must be inspected under a microscope. When viewed broadly, agricultural education encompasses all human endeavours including the collection, dissemination, and assimilation of knowledge in order to improve methods and comprehension of the procedures leading to scientific farming. Despite the fact that the current educational system produces graduates with the highest level of scientific knowledge, the required emphasis on the development of skills and attitudes, which needs to be strengthened, is missing. Only quality students could compete on the global market. As a result, it is essential that our education be skill- and business-oriented while also meeting market expectations.

Agricultural pathogens are a persistent source of issues for productivity and food safety due to their ongoing expansion and evolution. The numerous microorganisms connected to, or introduced into, agricultural systems, food, and related businesses provide advantages. The transnational movement of agricultural goods, trade regulations, industrial agricultural methods, and the potential for disease releases that are malevolent all result in new risks for agriculture. Research in microbiology is producing science and technologies that can assist agriculture in overcoming these difficulties. As biofertilizers, phytostimulators, bioherbicides, biopesticides, bioremediators, etc., beneficial microbes may be used in agriculture.

To enable the best utilization of beneficial microorganisms and maximum control of diseases, it is important to better understand the complex interactions between microbes and agricultural systems. An in-depth understanding of the sustainable use of microorganisms as an important tool in agriculture will be attempted in this curriculum. Reading this book, most possibly, basic knowledge of plant science, microbiology, biochemistry, and molecular biology will be improved.

Authors

Contents

1

Soil Organisms

Introduction

Millions of living organisms present in soil make it a living and a dynamic system. Under a microscope, it reveals a complex array of soil particles and pore spaces filled with air and water. In these pore spaces that plant roots and millions of organisms develop, which are ranging from sub-microscopic to macroscopic in size. The physical and chemical characteristics of soil determine the nature of the environment in which vast number of soil microorganisms especially soil microorganisms are found. Living organisms in soils include both macro (earthworms, rodents, termites etc.) and microorganisms (bacteria, fungi, actinomycetes, algae and protozoa). These organisms not only help in the development of soils but also carry out a number of transformations, facilitating the availability of nutrients to the plants.

The field of soil microbiology was explored during the very last part of 19th century. The establishment of the principal roles that microorganisms play in the biologically important cycles of matter on earth: the cycles of nitrogen, sulphur and carbon was largely the work of two men, S. Winogradsky (1856-1953) and M.W. Beijerinck (1851–1931). S. Winogradsky, the Russian scientist and regarded by many as the founder of soil microbiology, discovered nitryfiying bacteria (1890-91); described the microbial oxidation of H_2S and sulphur (1887); developed and contributed to the studies of reduction of nitrate and symbiotic nitrogen fixation; and, originated the nutritional classification of soil microorganisms into autochthonous (humus utilizers) and zymogenous (opportunistic) groups. Almost equally important was the work of M.W. Beijerinck, a Hollander, who isolated the agents of symbiotic (1888) and non-symbiotic aerobic (1901) nitrogen fixation. However, the greatest contribution of Beijerinck was a new and profoundly important technique: enrichment culture technique: to isolate and study various physiological types of various microorganisms from natural samples through the use of specific culture media and incubation conditions.

Classification of soil organisms

Although there is no single accepted classification for microorganisms, however the most expansive-Animalia, Plantae, Fungi, Protista and Monera kingdom have been proposed by Whittaker (1969).

The system of classification is based on three levels of cellular organisation which evolved to accommodate three principle modes of nutrition-photosynthesis, nutrient uptake by absorption and nutrient uptake by ingestion. Microorganisms are found in three of the above five kingdoms i.e. in Monera (bacteria, actinomycetes and cyanobacteria), Protista (micro-algae and protozoa) and Fungi (yeasts, mushrooms and molds).

Soil organisms can also be classified on the basis of requirements of molecular oxygen, temperature and mode of nutrition etc. Organisms that need oxygen for respiration and cannot survive without it are called obligate aerobes, and those which are aerobic but also adapt to grow under anaerobic environment using oxidised substances like NO_3, SO_4, CO_2 etc. As terminal electron acceptors in place of O_2 in respiration, are classified as facultative anaerobes. An example of such organisms of great importance in soil is the denitrifying bacteria. The organisms which grow in the absence of O_2 are called obligate anaerobes like Clostridia.

The organisms have specific optimum temperature requirements for their growth. Any change from the optimum temperature will not kill the organisms but will reduce their growth rate. Psychrophiles have optimum temperature for growth below 10 °C while the group of organisms with optimum temperature between 20-35 °C are termed mesophiles. This group is highly dominant and numerous in most of the cultivated soils of India. A temperature higher than 45°C favours development of thermophiles, commonly encountered in the compost pits.

Based on the mode of nutrition, soil organisms are classified into two groups - heterotrophs and autotrophs. Heterotrophs derive energy by oxidation of organic compounds while autotrophs derive their carbon from CO for cell synthesis. The group of autotrophs is further sub-divided into chemoautotrophs which get energy from the oxidation of simple inorganic compounds, and photoautotrophs, which derive energy from sunlight.

Macroorganisms in soil

The macroorganisms in soils include Acari, Collembola, Enchytracidae, Isoptera, Isopoda, Amphipoda, Diplopoda, Earthworm, Coleoptera, Mollusca, etc. A population or biomass estimation of these soil animals is highly difficult because they are not uniformly distributed in the soil and are highly mobile. Even though fewer in numbers than microorganisms, these organisms are very useful in soils as: (i) They help in the decomposition of organic residues in soil by mixing, churning or fragmentation as they eat on plant material, (ii) They form burrows and tunnels which increase soil aeration, drainage and turn in large amount of surface soil, (iii) With organic material as their food, the ingested soil in the guts of earth worms gets converted into worm casts called mull humus, and (iv) Some of these organisms like mites, termites and acrinemites feed on soil microorganisms including plant pathogens. Some of the soil macroorganisms which play an important role in soil fertility are discussed below.

Earthworms

Earthworms belong to the class Annelida and their several genera are found in soil. There are about 1800 species of earthworms in the World. Their length generally varies between 10-25 cm. Uncultivated grasslands and fallow pastures, having a generous amount of plant debris, have more number of earthworms. The total biomass of earthworms in soil ranges from 110-1100 kg/ha. Geophagus species of earthworms ingest material per day which is 5-36 times of their body weight. Casting rates of tropical earthworms are reported to be as high as 2600 t/ha/year. Earthworms do intimate mixing of organic matter with mineral matter which increases the stabilization of clay bound carbon, depending on soil type. Earthworm-worked soils generally have high porosity, increased water holding capacity, higher water infiltration rate, more water-stable aggregates and increased availability of plant nutrients. These organisms may also affect microbial population as they ingest microbes along with soil and organic matter. Alternatively, they increase the surface area and availability of organic matter for microbial action by mixing it thoroughly with soil.

Termites

Termites also influence soil properties and processes through four types of activities: (i) physical modification of soil profiles by constructing mounds, sheeting and foraging run ways; (ii) changes in soil texture emanating from movement of clay fractions from subsoil for constructions they make; (iii) changes in the nature and distribution of organic matter and plant nutrients through litter brought into nests which is digested by termites and decomposed by the microbes in situ (The termite-worked soil has higher CEC and exchangeable bases than surrounding soil); and (iv) The changes in soil drainage and moisture regimes upon constructing subterranean galleries. In intensively cultivated soils, the population of termites is low. However, forest and pasture lands have mound forming termites. Termites have cellulose decomposing microbes in their guts and thus their excreta has lower organic matter content than that of earth worms.

Plant Roots

Plant roots, which have immense role in soil formation, its fertility and productivity, can also be considered as one of the soil macroorganisms. Plant roots exercise tremendous influence on soil properties. However, being the primary source of organic matter in soil, plants and their roots supply food and energy to saprophytic soil organisms and maintain the soil biological activity. In addition, by exerting physical pressure on soil particles, they form many channels and capillaries, press the soil making cracks changing bulk density in microsites. In the long run, plant roots also influence soil texture but more often improve soil structure. By producing different chemicals as root exudates, roots influence the chemical and biological environment of soil around them. The term Rhizosphere was coined to encompass these influences. An environment under the influence of roots, the rhizosphere is chemically, biologically and probably physically also a different environment. Roots produce (i) exudates - chemical compounds leaking from roots; (ii) secretions - chemical compounds released through plant metabolic processes; (iii) mucilages - complex compounds produced by roots or through bacterial degradation; (iv) mucigels - gelatinous layers composed of mixture of mucilages and soil particles; and (v) lysates - compounds released from root cells through bacterial deg radation. These compounds at the soil root interfaces have a special kind of niche for the proliferation of microbes. Such niches are different from the root surface in contact with mucigel called rhizoplane.

Microorganisms in soil

Bacteria

Bacteria are the smallest and most numerous of the organisms present in soil. They are single cell organisms and their size varies between 1 micron to 10 micron in length. In spite of their small size, the bacterial biomass could be as high as 3500 kg ha of surface soil due to their large population. The bacteria in soil are of different shapes. Those with spherical cells are called cocci, rod-shaped cells are called bacilli and long spiral-shaped are termed spirilla. The number of bacteria is highly variable depending upon soil type, nature of crop cover and climatic conditions. Generally, soils with low organic matter and sandy texture have very low population. Some of the bacteria survive in soil under unfavourable conditions by forming spores which are resistant to prolonged dehydration and high temperatures. Direct examination of soil for bacteria shows that cocci and small rod-shaped bacteria are dominant in most of the soils. Some of the bacteria are surrounded by gummy capsules also.

The different bacterial genera commonly occurring in diverse soils are: *Pseudomonas, Arthrobacter, Clostridium, Bacillus, Achromobacter, Micrococcus* and *Agrobacterium*. The genus *Bacillus* has largest representation in soils in terms of species. Strictly anaerobic bacteria of the genus *Clostridium* occur in most soils, even under aerobic conditions. The occurrence of clostridia shows the presence of micro-anaerobic sites where these bacteria flourish. Aerobic soils receiving easily metabolizable organic matter also show anaerobiosis due to O_2 stress, created by rapidly developing aerobic bacteria which bring down partial pressure of O_2 and enrich soil atmosphere with CO_2 conducive for anaerobic growth. The presence of enterobacteria in soils, which strictly are not soil bacteria, is of great ecological significance, as it is indicative of fecal contamination. Their presence is a potential health hazard, particularly in growing crops whose edible plant parts come in direct con tact with soil. Thus, enteric pathogens may get into food chain and cause human diseases.

Bacteria due to their large number and very rapid rate of multiplication play a very significant role in carrying out various biochemical reactions controlling availability of plant nutrients. The processes of N_2 fixation, phosphate solubilization, organic matter decomposition and synthesis of humus, nitrification and denitrification, protein decomposition and ammonification, etc. lead to transformation of various macro- and micro-nutrients in soil and help in plant nutrition.

Certain free-living soil bacteria can reduce the atmospheric nitrogen to ammoniacal form, thus enriching soil with plant available nitrogen. These bacteria belong to the genus *Azotobacter, Azotomonas* and develop freely in soil which are relatively high in organic matter. The genus *Beijerenckia* and *Derxia* are unique; they can fix nitrogen in soils which are highly acidic (pH 4.0). *Clostridium* fixes nitrogen under anaerobic conditions and can be of economic significance in flooded soils. Besides these free-living bacteria, there is another group belonging to the genus *Rhizobium* which fixes nitrogen in symbiosis with the legumes forming tiny out-growths on roots called legume root nodules, the sites of nitrogen fixation. *Azospirillum*, a spiral bacterium and others be sides fixing nitrogen as free-living can also fix nitrogen in association with plants in rhizosphere entering into roots without forming nodule and the process is called "associative nitrogen fixation".

The microbiological oxidation of ammonium to nitrate is one of the most important trans formations taking place in soil which helps plant nutrition since nitrogen is predominantly taken up as NO, ion. This transformation is mediated by a group of bacteria called 'nitrifying bacteria'. It is a two-stage process wherein firstly ammonium is transformed into nitrite by bacteria belonging to the genus Nitrosomonas and in the second stage bacteria of the genus Nitrobacter convert nitrite to nitrate. As the two processes take place simultaneously, there is no accumulation of nitrite as an intermediate. These bacteria are aerobic in nature. Hence, supply of oxygen is essential for this process. When oxygen supply is restricted, the activity of nitrifying bacteria goes down, low ering the nitrification rate. Any cultural practice which improves aeration results in enhancing nitrification. The optimum pH for these bacteria is between 6.5 and 7.5. Therefore, under acidic conditions nitrification is slow and is nearly negligible below pH 5.0.

Actinomycetes

Taxonomically actinomycetes are like bacteria which possess aerial hyphae like fungi. These organisms share characteristics of both bacteria (cell size, structure and mode of multiplication) and fungi (branching). They are in an evolutionary phase between the two. They are next to bacteria in numbers and are fairly widely distributed in soils. They are more common in dry soils and in undisturbed pastures and grasslands. Like bacteria, they are more common in neutral to slightly alkaline soils.

They are aerobic organisms and therefore their number is less in lowlands. As they can withstand drought conditions very well, they occur more frequently in soils undergoing dry spells. The species more commonly encountered in soil belong to genera *Streptomyces*, *Micromonospora*, *Nocardia* and *Thermoactinomyces*. Many of the soil actinomycetes are known to produce antibiotics and have been commercially exploited to produce antibiotics of clinical importance. However, their role in biological equilibrium under natural soil conditions is doubtful as concentration of antibiotics produced by them seldom reaches concentration levels to inhibit the susceptible organisms.

Actinomycetes are nutritionally heterotrophic and are highly adaptive to degrade wide range of organic substances, particularly those which are difficult to be decomposed by other soil organisms. They are also slow growing organisms. Hence, they cannot compete with bacteria and fungi for the available nutrients and thus grow on substrates that are relatively more resistant to attack of bacteria and fungi. These organisms start functioning when easily decomposable fractions like sugars, starches, etc. are used up by bacteria and fungi and the more difficultly decomposable substances accumulate as they face less competition from other organisms. The role of actinomycetes in the synthesis of humus is well known. They are reported to produce a number of colour pigments contributing dark colour to soil humus. Actinomycetes population is also much larger in compost pits as they can with stand high temperatures. Some of them are pathogens causing plant diseases such as potato scab, etc.

Fungi

Fungi are filamentous organisms with much larger cell width than actinomycetes. The filaments are called hyphae and the network of hyphae collectively arable soils is less than that of bacteria, their biomass is more than bacteria being larger in size. Soil fungi can grow in wide range of soil pH but their population is more under acidic conditions because of severe competition with bacteria at neutral pH. A majority of fungi are aerobic and prefer to grow at optimum soil moisture. The contribution of these organisms in biochemical transformations under excessive moisture is negligible. Fungi are capable of sending hyphae into anaerobic zone in soil sustaining oxygen supply from the hyphae which develop in oxygen rich environment and hence the fungi growing in surface soil may send hyphae into deeper soil layers not sufficiently aerated.

Most of the soil fungi can be divided into three classes: (i) Phycomycetes, (ii) Asco mycetes, and (iii) Fungi Imperfecti. Phycomycetes are characterized by non-septate hyphae and spores are borne in sporangia. In Ascomycetes, the hyphae are divided by septa and a definite number of spores (usually eight), called ascospore, are borne in a sac like structure ascus. Fungi Imperfecti, which are most frequent in soils, multiply by means of asexual spores called conidia. The genera most frequently encountered are *Pythium, Rhizopous, Mucor, Cunninghamella* (Phycomycetes); *Chaetonium* (Ascomycetes); *Aspergillus, Penicillium, Verticillium, Alternaria, Fusarium* (Fungi Imperfecti). These fungi grow profusely in soils to which crop residues and organic manures have been added recently.

Fungi are heterotrophs and therefore derive nutrition from either living plant tissues as a parasite, causing plant diseases or dead tissues as a saprophyte. Saprophytic fungi perform a very important function in the decomposition of organic matter, particularly plant residues. Plants have rigid cell walls and low protein content which make them resistant to the action of slowly diffusing hydrolytic enzymes. Being filamentous, fungi are able to penetrate the rigid plant tissues due to the pressure exerted by the hyphal tips. The hydrolytic enzymes produced inside tissues cause local softening of the tissues, resulting in increased diffusion of enzymes produced at the surface.

Both bacteria and fungi then can act in tandem where bacteria move along with the fungal hyphae invading internal tissues, thus initiating de composition of internal tissues by bacteria.

Some fungi form a symbiotic association with roots of higher plants facilitating up take of plant nutrients, particularly of those which are less mobile. This association is known as "mycorrhizal association". There are two types of mycorrhizal association:

(i) Ectotrophic mycorrhizae where the fungus forms a mantle or sheath around the root surface (called Hartig net) and where the mycelium develops intercellularly. The fungi which form this type of association are species of *Boletus, Amenita*, etc. (ii) Endomycorrhizae where the fungus develops intracellularly in the root without forming hartig net. In this association, the penetration of roots cells is characterized by the formation of terminal spherical structure called vescicles which contain oil drop lets and phosphorus. This type of mycorrhiza is called "vescicular arbuscular mycorrhizae' (VAM) and is of agricultural significance particularly in P-deficient soils where the phosphorus in the vescicles diffuse out into the cytoplasm and is taken up by the plant. Fungi belonging to the genera *Glomus, Endogene* form this association. The beneficial effect of these fungi on nutrient uptake has been attributed to three factors: (i) Increased absorption of available nutrients from soil as the fungus changes root morphology which results in the larger root surface available for nutrient absorption. Fun gal filaments also act as absorption surface, (ii) Increasing the nutrient availability by solubilizing insoluble nutrients like P which thus become available to plant, and (iii) Increasing the nutrient mobility due to faster intracellular nutrient mobility and mobilizing nutrients from the soil mass not visited by the root system but traversed by the mycorrhizal hyphae.

Algae

Soil algae are chlorophyll containing organisms. They are autotrophic, therefore their development is not restricted by organic carbon supply. They are abundant in habitats exposed to light and have sufficient moisture. They are unicellular, filamentous or form colonies from single cells.

Soil algae are classified on the basis of the colour (pigments) as: (i) Cyanophyta (blue green), (ii) Chlorophyta (grass green), (iii) Xanthophyta (yellow green), and Bacilliriophyta (golden brown). Blue green algae, also known as 'Cyanobacteria', are most important from agricultural point of view because they fix atmospheric nitrogen and contribute towards the nitrogen economy of soils, particularly in rice cultures of tropics.

Blue green algae have been reported to fix about 20 kg N per ha. Some of the important genera dominant in rice fields are *Anabaena*, *Nostoc*, and *Tolypothrix*. The agricultural significance of blue green algae in the nitrogen economy of paddy soils is of immense importance. Submerged soil conditions in the rice field provide an ideal environment for the growth of algae as the soils seldom dry out. These organisms not only help in the N-nutrition of rice plants but are also reported to synthesize plant growth hormones and thereby help plant growth. As they have photosynthesizing capacity, they release O, which support aerobes in flood water. The beneficial effect of algal inoculation has been widely demonstrated in different types of soils and climatic conditions.

A nitrogen fixing algae, *Anabaena azollae*, forms a symbiotic association with a fresh water fern, Azollae and fixes nitrogen. The azolla biofertilizer not only contributes to wards soil nitrogen but also adds organic mat ter. Algae are also found in association with fungi called lichens. In this association, the algae provide carbohydrates, produced through photosynthesis, to the fungi and the fungal symbiont provides mineral nutrition and regulates water supply. Lichens are distributed worldwide on rocks, soils, buildings, foliage, tree trunks, etc. Usually lichens are amongst the first colonizers of barren rocks and initiate soil formation.

Protozoa

Soil protozoa are single cell organisms belonging to animal kingdom and are larger in size than most microorganisms found in soil. The life cycle of protozoa consists of an actively growing phase when it multiplies, and a resting phase when under adverse environmental conditions like low moisture and high temperature, vegetative cell covers itself with a thick coating which is called 'cyst'. The protozoa are classified on the basis of mode of locomotion. Some move by long whip like structures called flagella and others by short hair like appendages called cilia; and still others by internal protoplasmic movement forming flex ible temporary organ called pseudopodia or false foot.

In fertile soils, the number of protozoa may be up to 1 million per gram. Soils which are moist contain a large population of these organisms. As most of the soil protozoa are saprophytic, their population is high in soils with high organic matter. Some of the protozoa, however, directly feed on bacteria present in soil. Therefore, it was once thought that high protozoa population is detrimental to soil fertility because they reduce population of useful bacteria in soils. It is now considered that protozoa do not reduce bacterial population permanently, and moreover, soil organisms other than bacteria are also involved in plant nutrient availability. Protozoa mainly influence the organic cycle. Their predatory nature on bacteria contributes to the turnover of available nutrients and prevents immobilization of nutrients by keeping a check on bacterial population.

Nematodes

Amongst the microfauna, the nematodes are next in abundance to protozoa. Because of their narrow long bodies, they are also called thread worms. They may be either saprozoic feeding on decaying organic matter or parasitic on living plants. Due to less number in soils, they are not of much significance in organic matter decomposition. However, parasitic nematodes cause plant diseases and thus gain significance in some localized pockets due to high population. By infesting plant roots some nematodes form characteristics knots. In India, vegetable crops are particularly susceptible to nematode infestation. They are controlled by chemical fumigants. Amendment of soils with some non-edible cakes from neem and karanj and saw dust is also reported to control these parasites. Nematode trapping fungi parasitise on them in soil. These organisms otherwise have little role in soils.

Viruses

Viruses are ultramicroscopic organisms smaller than bacteria which cannot be seen by an ordinary light microscope. They parasitise on animals, plants and microorganisms. Viruses parasitising bacteria are known as bacteriophages. When bacteriophages develop rapidly in a soil, they may kill useful soil bacteria and make the soil sick. A large scale failure of nodulation in legumes is indicative of the presence of bacteriophages of the specific rhizobia. Sometime a soil may act as a sink for viruses causing human diseases. For example, Hepatitis virus survives longer in soil than in any other habitat. Similarly, several enteric viruses can be traced to soil where they remain dormant for long periods and become active after they infect a suitable host. Soil viruses do not take part in nutrient transformations due to their parasitic nature. However, they may affect nutrient cycles indirectly by parasitising bacteria responsible for these cycles such as nitrifying bacteria, nitrogen fixing bacteria, cellulose hydrolysers, etc.

Objective type questions

A. Fill up the blanks with suitable words

i. ____________________ is an example of symbiotic nitrogen fixing bacteria.

ii. Organisms that need oxygen for respiration and cannot survive without it are called ____________________.

iii. The group of organisms with optimum temperature between ______________ °C are termed mesophiles

iv. Obligate aerobes are aerobic but also adapt to grow under anaerobic environment using oxidised substances like _______ & ________.

v. _________________, the Russian scientist is regarded by many as the founder of soil microbiology, discovered nitryfiying bacteria (1890-91)

vi. _________________, the scientist has discovered nitryfiying bacteria during the year 1890-91.

vii. ______________________, a Hollander, who isolated the agents of symbiotic (1888) and non-symbiotic aerobic (1901) nitrogen fixation

viii. Five kingdom classification of microorganisms was proposed by ___________________ in the year____________.

ix. Photoautotrophs, which derive energy from __________________.

x. Earthworms belong to the class ____________

B. Multiple choice questions

i. Which of the following type of soil has highest number of earworm count in soil -

a. Forest soil

b. Garden soils

c. Continuously cultivated soil

d. Uncultivated grasslands and fallow pasture

ii. The total biomass of earthworms in soil ranges from -

a. 11-110 kg/ha   b. 50-500kg/ha

c. 110-1100 kg/ha   d. None

iii. Earthworm-worked soils generally have -

a. High water stable aggregates   b. high porosity

c. increased water holding capacity   d. All the above.

iv. Bacteria are single cell organisms and their size (length) varies between

a. 1 μ to 10 μ   b. 10 μ to 100 μ

c. 100 μ to 1000 μ   d. None

v. Rod-shaped bacterial cells are called as -

a) Bacilli b) Cocci

c) Spirilla d) None

vi. Which of the following bacterial genus is strictly anaerobic in nature -

a) Clostridium b) Bacillus

c) Achromobacter d) Micrococcus

vii. Dark brown coloration of humus is due to the presence of ------------------ in soil.

a. Actinomycetes b. Bacteria

c. Fungi d. Termites

viii. Which of the following microorganisms directly feed soil bacteria -

a. Actinomycetes b. Bacteria

c. Fungi d. Protozoa

ix. Which of the following organism could fix atmospheric nitrogen -

a. Earthworms b. Azolla

c. Fungi d. Termites

x. Fungi generally prefers which of the following types of soil reaction -.

a. Acidic b. Saline

c. Alkaline d. Neutral

Answer keys:

A. Fill in the blanks

i. *Rhizobium* sp., **ii.** Obligate aerobes, **iii.** 20-35 °C, **iv.** NO_3 & SO_4, **v.** S. Winogradsky, **vi.** S. Winogradsky, **vii**. M.W. Beijerinck, **viii**. Whittaker in 1969, **ix.** Sunlight, **x.** Annelida

B. Multiple choice questions:

i. **d** ii. **c** iii. **d** iv. **a** v. **a** vi. **a** vii. **a** viii. **d** ix. **b** x. **a**

C. Descriptive type questions

i. Classify microorganism according to five kingdom classification proposed by Whittaker.

ii. Define mycorrhizae. Classify mycorrhizal association with two examples for each class.

iii. Write in brief the functions of soil macroorganisms in decomposition of soil organic matter.

iv. How earthworms could influence soil fertility?

v. Classify bacteria according to their shapes. Write in brief the importance of beneficial soil bacteria in improving soil fertility.

2

General Properties and Classification of Bacteria

Introduction

The microorganisms exist as single cell or cluster of cells capable of orderly growth, metabolism and reproduction in two identical daughter cells. These includes cellular organisms {protozoa, algae and fungi (Eukaryotes), bacteria and archaea (Prokaryotes)} and viruses (non cellular entities). The study of microorganisms contribute to understanding of :

1. Basic biological sciences - It acts as a tool for providing the basic information for life processes. The discovery of DNA, RNA, proteins and enzyme functions all came from microbial study.
2. Applied biological sciences - Medicine, agriculture, industry, plant and animal pathology, immunology, plant and animal nutrition, biotechnology all have origin from the understanding of the cell function and metabolism of microorganisms.

The common features of all biological systems are as follows:

i) All organisms share common chemical composition.

ii) All organisms perform certain chemical activities.

iii) All organisms share common physical structures.

Chemical composition

Three types of complex organic macromolecules present in organisms are: proteins, deoxyribonucleic acid (DNA) and ribonucleic acid (RNA). Proteins are biological catalysts or enzymes. These are involved in different types of biological reactions. DNA is the genetic material. RNA is responsible for protein synthesis.

Metabolism

It involves the processes required for growth and to generate energy necessary for their activities. The metabolic functions involve **catabolism** - breaking of preformed inorganic or organic molecules for energy generation and formation of precursor molecules, and **anabolism** - synthesis of biomolecules with use of energy and precursor molecules for enzyme synthesis, maintenance, growth, production of macromolecules and reproduction. Catabolism and anabolism are associated with enzyme specific functions. Metabolism is also governed by genetic variations in different microorganisms.

Physical structures

All organisms are organized into microscopic units known as cells. As a result of their cellular organization, growth results in cell division, with an increase in total number of cells. There are two types of cellular organization among microorganisms:

i. Unicellular organization: It consists of small microscopic single cells. It is most common in bacteria, protozoa and yeasts.

ii. Multicellular organization: It consists of many cells attached to one another in a characterized fashion. Multicellular organization arises initially from single cells. It is most common in algae and fungi.

Unicellular microbial cells have independent existence whereas multicellular organisms with cell differentiation does not have independent existence. Viruses do not have cellular existence and their own metabolisms. Their metabolic activity and reproduction is host dependent as they are intracellular parasites.

The cell is the fundamental unit in microorganisms, each separated from the other by a cell membrane (and perhaps a cell wall except in case of animals). The cytoplasmic (cell) membrane is the barrier in all cells separating inside from outside. Nutrients are selectively taken in and waste products of cell pass out. The cell membrane is very thin, flexible and structurally weak. Any leakage or damage to cell membrane results in death of the cell. Usually the cell membrane is covered with a rigid stronger outer layer, called the cell wall. Plants and microbial cells have rigid cell walls but animal cells are devoid of cell walls.

Inside the membrane is a complicated mixture of substances called the cytoplasm. The inner cytoplasmic portion contains genetic, structural and cytoplasmic constituents essential for growth, metabolism and multiplication.

The cytoplasm contains water-containing macromolecules, ribosome, cellular precursor molecules, the enzymes and inorganic nutrients. The macromolecules - proteins, lipoproteins, lipids, polysaccharides and nucleic acids are required for growth, reproduction and metabolic functions. The diversity of microorganisms are due to evolutionary trends arises during the long periods of their existence on the Earth's surface from the universal ancestor.

Microbial classification

Earlier biologists recognized two kingdoms of living organisms based on simple characters like motility and ability to carry out photosynthesis. These kingdoms were (i) The plant kingdom and (ii) The animal kingdom.

Table: Major differences between the plant kingdom and animal kingdom

	Characteristics	Plant kingdom	Animal kingdom
A.	**Functional characters**		
1.	Energy source	Light (photosynthesis)	Organic compounds as energy source (non-photosynthetic)
2.	Motility	Absent (immotile)	Present (motile)
3.	Carbon source	Carbon dioxide	Organic compounds
4.	Growth factors	None	Complex
B.	**Structural characters**		
5.	Cell walls	Present	Absent
6.	Chloroplasts	Present	Absent
7.	Mode of growth	Open (continues throughout life)	Closed (fixed size and form as an adult)

i) Algae: Photosynthetic and immotile, hence classified in plant kingdom.

ii) Fungi: Non-photosynthetic (AK) and immotile (PK). Therefore, these fungi were classified in animal kingdom.

iii) Protozoa: They possess both the characters of the animal kingdom i.e. motile and non-photosynthetic. Hence, they were classified in animal kingdom.

iv) Bacteria: Though possessed both the characters of animal kingdom i.e. motile and non-photosynthetic but these were classified in plant kingdom on the basis of rigid cell wall.

This type of overlapping characters posed problem in classification. Haeckel (1866) suggested a third kingdom known as Protista. All the groups of microorganisms like protozoa, algae, fungi and bacteria (members of which share the properties of both animal and plant kingdom) were included in this kingdom.

This kingdom protista was divided into two groups:

a. Higher protists or eukaryotes which include fungi, algae and protozoa.

b. Lower protists or prokaryotes which include bacteria and blue green algae (cyanobacteria).

Eukaryotic cells and Prokaryotic cells

1. Eukaryotic cells: Eu means true; karyotic means containing nuclear material. hus Eukaryotic cell means cells containing true nuclear material. Cells are most complex in nature and are commonly found in algae, fungi, protozoa, plants and animals.

2. Prokaryotic cells: Pro means like; karyotic means containing nuclear material. Thus prokaryotic cell means cells containing nuclear-like material. In prokaryotic cell, nucleus is not present but material similar to nucleus is present. Prokaryotic cells are found in bacteria and cyanobacteria (blue green algae).

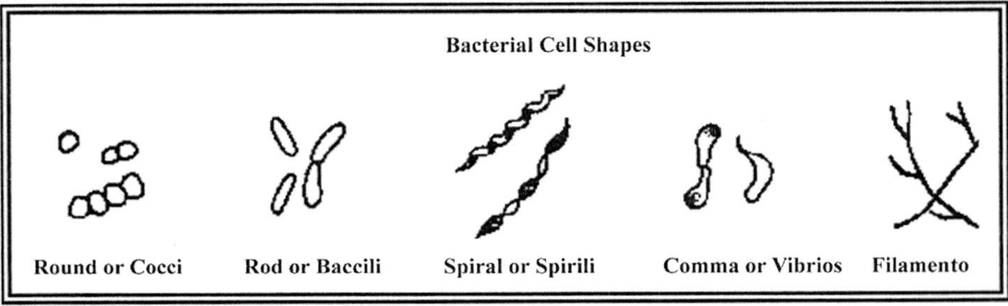

Fig. Bacterial cell.

Table: Major differences between Eukaryotic cells and Prokaryotic cells.

Component	Eukaryotic cells	Prokaryotic cells	Functions
Group of organisms found	Algae, fungi, protozoa, plants and animals	Bacteria	-
Cell size	5.0 µm or more in diameter	0.5 to 1.0 µm in diameter 3.0 to 6.0 µm in length	-
Capsule	Absent	Present	Imparts pathogenicity in pathogens
Cell membrane	Sterols present	Sterols absent: lipids and proteins are present	Osmotic barrier and permeability
Nucleus	Nuclear membrane present; contains chromosome, DNA	Nuclear membrane absent; DNA found disperse in cytoplasm	Genetic material

Component	Eukaryotic cells	Prokaryotic cells	Functions
Ribosomes	70S in organelle and 80S in cytoplasm	70S	Protein synthesis
Mitochondria	Present	Absent	Site of respiration
Chloroplast	Present (absent in yeast and protozoa)	Absent (present in cyanobacteria and green sulphur bacteria)	Photosynthesis
Endoplasmic reticulum	Present	Absent	Site of ribosome attachment
Golgi bodies	Present	Absent	Packaging of material synthesized in ER and its transport to parts of cell
Mesosomes	Absent	Present	Participate in septum formation
Movement	Present except in plants	Present	Locomotion
Cytoplasmic streaming	Present	Absent	Translocation of solids
Anibiotic sensitivity			
Penicillin	Resistant	Sensitive	Peptidoglycan synthesis is inhibited
Nystatin	Sensitive	Resistant	Inhibits incorporation of sterols in cell membrane
Chloramphenicol	Resistant (but sensitive at higher concentration)	Sensitive	Inhibits protein synthesis on 70S ribisome
Streptomycin	Resistant	Sensitive	Inhibits protein synthesis on 70S ribisome
Cycloheximide	Sensitive	Resistant	Inhibits protein synthesis on 80S ribisome

Classification of organisms

Microbiologists use the Linnean system of binomial nomenclature to name the microorganisms. An organism's name is made of genus and species.

Species can be defined as group of similar individuals that are sufficiently different from other individuals, to be considered as a recognized taxonomic group. e.g. *Rhizobium meliloti*, *R. leguminosarum* and *R. fredii*. These bacteria are capable of making nodules but on different hosts i.e. alfalfa, pea and soybean.

A collection of species that share a major property (or properties), making them a distinct grouping, permits the group to be considered as a genus. For example, the genus Rhizobium is fast-growing, Gram positive, aerobic rods with the capability to make nodules on legume roots.

Historically, microorganisms are classified on the basis of taxonomic features and characters which include structure, morphology, staining reactions and physiological parameters including ability to grow on particular carbon source/carbohydrate, growth on particular temperature and oxygen levels etc. However these features are phenotypic (based on physical characteristics) rather than phylogenetic (based on genetic relationships). The technology of particular sequencing has introduced a totally new way of determining relationships among organisms. Phylogenetic trees showing relationships among organisms are constructed directly from comparisons of informational macromolecules, such as ribosomal RNA (rRNA) genes, occurring in living cells.

Bacteria and Archaea

These domains comprise a remarkably diverse group of single-celled prokaryotic organisms which inhabit the soils of every terrestrial ecosystem from the warm, moist, densely vegetated soils of tropical rain forests to the deserts, grasslands, forests of temperate regions and the frigid tundra of the high altitudes. Despite their small size, the bacteria and archaea exhibit greater metabolic capabilities than in any other groups of organisms and play crucial role in biological transformation of nutrients in the soil, organic matter decomposition, remediation of contaminated soils, in mutualistic interactions with plants and in plant diseases.

In a fertile soil, the population of bacteria is 10^8 to 10^9 cells/gram of soil. They are unicellular and microscopic. Bacteria are so small that a special unit of measurement (micron) is used while describing their size. They constitute one of the largest groups. These microorganisms may be rod shaped, round or spiral in shape. Some bacteria are autotrophic and many are heterotrophic. Bacteria are the most abundant group, usually more numerous than any other combined groups of soil microorganisms. Although the bacteria are smaller in size as compared to other microbial groups, but they dominate the microbial transformation of plant nutrients because of their rapid growth rate and vigorous decomposition ability.

In Bergey's Manual, bacteria have been classified in four divisions based upon the type of wall that the cells possess.

Division I. Gracilicutes: Prokaryotes with a Gram negative cell wall.

Division II. Firmicutes: Prokaryotes with a Gram positive cell wall.

Division III. tenericutes: Prokaryotes that have no cell wall, commonly called mycoplasmas.

Division IV. Mendosicutes: Prokaryotes with walls that do not contain the bacterial polymer peptidoglycan (Archaea).

Current phylogenetic classification schemes place division I-III in the domain bacteria and division IV in the archaea. The archaea include microorganisms that grow in harsh environments (extreme halophiles and thermophiles) and strictly anaerobic methanogens, which can reduce carbon dioxide to methane gas.

Differences between the domain bacteria and archaea

Character	Bacteria	Archaea
Lipids in membrane	Ester-linked, straight chain fatty acids	Ether-linked, branced chain aliphatics
Cell walls	Peptidoglycan-muramic acids	Variety, no muramic acid; Pseudomurein
Transfer RNA (tRNA)	Thymine present	Thymine absent
Ribosome response to Chloramphenicol, Kanamycin	Sensitive	Insensitive
Antisomycin	Insensitive	Sensitive
DNA-dependent RNA polymerase		
Number of enzymes	One	Several
Structure	Simple subunit	Complex subunit
Rifampicin sensitivity	Sensitive	Insensitive
Response to diphtheria toxin	Insensitive	Sensitive

Winogradsky (1925) divided the soil microorganisms in general and the bacteria in particular into two main categories:

1. Autochthonous or indigenous microorganisms

These are the true residents of the soil. Their population is always uniform and they derive their nutrition from soil organic matter consisting of resistant constituents of plant residues and components of dead microbial cells as nutrients. Due to less readily availability of nutrients, such microorganisms grow slowly and their abundance is not subjected to marked fluctuations of the environmental conditions. They may have resistant stages to survive for long durations without being active metabolically. Sometimes, these microorganisms proliferate and participate in biochemical functioning of the community e.g. *Bacillus*, *Arthrobacter* and *Pseudomonas*.

2. Allochthonous or zymogenous (fermentative) microorganisms

These microorganisms get their entry in the soil with precipitation, diseased plant tissues, animal manures or sewage sludge. Their normal population in the soil is low. However, when specific substrates are added into the soil, the number of zymogenous bacteria increases. When the added substrates are exhausted, their population gradually declines. e.g. cellulose decomposers, ammonifiers and nitrifying bacteria.

Basis of classification in lower protists (Prokaryotes)

The lower protists are further classified based on the characters like (i) Gram staining, (ii) Cell morphology, (iii) presence of endospore, (iv) oxygen requirement, (v) temperature requirement for growth, (vi) base composition of DNA, (vii) Serology, (viii) Carbon and energy source and (ix) rRNA phylogenetic approach.

1. Gram staining: Gram staining was discovered by Cristian Gram in 1884. Based on this staining, bacteria are divided into Gram-positive and Gram-negative bacteria. Gram-positive bacteria are those which retain the violet colour of primary stain (crystal violet) even after decolorization with alcohol. e.g. *Bacillus, Clostridium* and *Streptococcus.*

Gram-negative bacteria are those which loosen the colour of crystal violet stain after decolorization with alcohol and get stained with secondary stain safranine. e.g. *Rhizobium, Pseudomonas* and *Azotobacter.*

Sl.No.	Gram-positive	Gram-negative
1.	Walls of Gram-positive bacteria contain peptidoglycan accounting to 40 to 90% of the wall dry weight.	Walls of Gram-negative bacteria have a low peptidoglycan content seldom exceeding to 05 to 10% of the wall dry weight.
2.	Cell wall is single homogeneous layer, 10-80 nm in width.	Cell wall is composed of two distinguishable layers; Inner membrane (2-3 nm wide) and outer membrane (7-8 nm wide)
3.	Peptidoglycan matrix is covalently linked to polysaccharide and polyphosphate polymers known as Techoic acids	Techoic acids and polysaccharides are absent
4.	Proteins are absent from the walls of most Gram-positive bacteria	Various proteins and lipoproteins are present in both the membranes
5.	Lipopolysaccharides are absent	Lipopolysaccharides are present
6.	Walls of Gram-positive bacteria contain no lipid except few bacteria (i.e. Corynebacteria, Mycobacteria and Nocardias)	Phospholipids account for 20-30% of the dry weight of the bacterial cell membranes

2. Morphology

The morphological characters of bacteria include shape, size and flagellation.

i) Shape: There are three principal shapes in bacteria;

a. Cocci (Coccus - Singular) round and spherical cells. e.g. Micrococcus, Spreptococcus, Diplococcus and Staphylococcus.

b. Rod shaped bacilli (Bacillus - Singular) rectangular or rod shaped. e.g. *Bacillus, Rhizobium, Pseudomonas* and *Xanthomonas*.

c. Spirilla (Spirillum - Singular) spiral shaped. e.g. *Azospirillum, Spirochaetes* and *Herbaspirillum.*

ii) Size: Size of bacterial cell varies from species to species. Generally cocci measure from 0.5 μm to 1.0 μm in diameter and rods measure upto 3.0 to 6.0 μm in length and 0.5 μm in width.

iii) Flagellation: The locomotive organ in bacteria is known as flagella and the arrangement of flagella on the bacterial cell is called flagellation. Bacteria without flagella are called atrichous. There are different types of flagellation.

a) Polar: When single flagellum (monotrichous, e.g. *Vibrio*) or a tuft of flagella (cephalotrichous) are present on the one end of the cell. e.g. Pseudomonas.

b) Bipolar: When single flagellum (Amphitrichous, e.g. *Alcaligenes faecalis*) or a tuft of flagella (lophotrichous) are present on both the ends of the cell. e.g. Azospirillum.

c) Peritrichous: When flagella are arranged all over the cell. e.g. Bacillus.

3. Presence of endospores

Some bacteria produce specialized structures known as endospores to overcome unfavourable conditions. Endospores differ from vegetative cells in their chemical composition and other properties like resistance to heat up to 120^0C, toxic chemicals, stains, UV light and ionizing radiations etc. Examples of endospore producing bacteria are Bacillus, Clostridium, Sporolactobacillus, Desulfatomaculum and Sporosarcina. Endospores are observed in the bacterial cell after staining with malachite green and safranine stain. The green coloured spores are found enclosed in the pink-coloured vegetative cell.

4. Relation to oxygen

i) Aerobes: Those bacteria which need oxygen for their growth and respiration. e.g. *Bacillus, Rhizobium, Pseudomonas* and *Azotobacter.*

ii) Anaerobes: Those bacteria which grow in the absence of oxygen using inorganic compounds (e.g. NO_3 or SO_4) for respiration. These bacteria use fermentative pathways for generating energy rich compounds such as ATP. e.g. *Clostridium, Methanobacterium* and *Methanococcus.*

iii) Microaerophilic: These organisms use oxygen as the terminal electron acceptor with a strict respiratory type of metabolism. However, they grow best at partial pressures of oxygen considerably below the atmospheric O_2 concentration (i.e. 21%). e.g. *Azospirillum sp.* grow best under 1% O_2 concentration and *Campylobacter fetus* grows best under 6% O_2 concentration.

iv) Facultative anaerobes: These bacteria have diversity in respiration, using mainly O_2 for respiration but in case of partial anoxygenic environment, they could utilize NO3, SO_4 or other inorganic compounds for respiration. They could also derive energy from fermentative pathways. *Escherichia coli, Salmonella, Proteus, Shigella.*

v) Aerotolerant anaerobes: These bacteria grow under both aerobic and anaerobic conditions but do not shift from one mode of metabolism to another as the conditions change. They obtain energy exclusively by fermentation. They do not form toxic products by using O_2 as terminal electron acceptor. Therefore, aerotolerant anaerobes are not poisoned by O_2 in the same manner as obligate anaerobes. e.g. *Lactobacillus.*

5. Nutritional grouping of bacteria

On the basis of carbon uptake, the bacteria are grouped as autotrophic and heterotrophic.

i) Autotrophs (syn. Lithotrophs): utilize inorganic CO_2, HCO_3 or CO_3 as sole source of carbon.

ii) Heterotrophs (syn. Organotrophs): using organic carbon compounds as carbon source.

On the basis of energy derivation, they could be grouped as:

i) Phototrophs: using energy from sunlight for electron activation to generate ATP. For example,

a. Blue green bacteria (Cyanobacteria). e.g. Anabaena.

b. Green sulphur bacteria. e.g. *Rhodospirillum, Rhodomicrobium.*

c. Purple sulphur bacteria. e.g. *Chromatium, Thiospirillum.*

ii) Chemotrophs: utilizing energy from oxidation-reduction reactions of inorganic substrates (i.e. NO_3, SO_4, Fe^{++}, H_2 and other inorganic compounds) for generating ATP. For example, *Nitrosomonas, Nitrobacter, Thiobacillus, Gallionella* and *Hydrogenomonas.*

On the basis of carbon uptake and enrgy derivation, the bacteria are grouped into four categories:

i) Photoautotrophs: A bacterium utilizing CO_2 as a sole source of carbon and generating energy from photosynthetic process is photoautotrophic. e.g. *Chromatium.*

ii) Photoheterotrophs: A bacterium utilizing organic carbon source and generating energy from photosynthetic process is photoheterotrophic. e.g. *Rhodopseudomonas.*

iii) Chemoautotrophs: A bacterium utilizing CO_2 as a sole source of carbon and generating energy from oxidation-reduction reactions of inorganic compounds is chemoautotrophic. e.g. *Nitrosomonas, Thiobacillus* and *Desulphovibrio.*

iv) Chemoheterotrophs: A bacterium utilizing organic compounds as a source of carbon and generating energy from oxidation-reduction reactions of organic compounds is chemoheterotrophic. e.g. *Bacillus, Arthrobacter* and *Pseudomonas.*

6. Temperature requirement

Bacteria are grouped into three classes on the basis of temperature requirement for optimal growth.

i) ***Mesophiles*****:** Bacteria that grow optimally at temperatures in the range of 15 to 35^{0}C. Most bacteria are mesophiles.

ii) ***Thermophiles*****:** Some bacterial species survive and often thrive at temperatures in excess of 40 to 50^{0}C and as high as 100^{0}C. Thermophilic bacteria produce heat stable enzymes and structural proteins that do not denature at elevated temperatures. e.g. Taq polymerase isolated from *Thermus aquaticus* is used in PCR for DNA amplification (72^{0}C).

iii) ***Psychrophilic bacteria*****:** These bacteria are adopted to grow optimally at temperatures below 15^{0}C. These psychrophilic bacteria often synthesize large quantities of unsaturated fatty acids that maintain the fluidity of the cytoplasmic membrane at low temperature.

Category	Temperature optima	Capacity to grow at temp.
Psychrophiles	Below 15^{0}C	-3 to 20^{0}C
Mesophiles	15 to 35^{0}C	15 to 45^{0}C
Thermophiles	45 to 65^{0}C	40 to 100^{0}C

Some bacteria persist under unfavourable conditions by the formation of endospores. These endospores often endure in adverse environments because of their resistance to both prolonged desiccation and to high temperatures.

7. Base composition of DNA

Nucleic acids of bacteria contain DNA and RNA. DNA is composed of deoxyribose sugar, nitrogenous base and phosphate moiety. Guanine and adenine are purine bases whereas cytosine and thymine are pyrimidine bases. In RNA, the thymine base is replaced by uracil and the sugar is ribose. The amount of adenine is equal to amount of thymine and they pair with two hydrogen bonds. The amount of guanine is equivalent to the amount of cytosine and they pair with three hydrogen bonds. The amount of GC and AT varies widely with different species of bacteria.

$$\text{Relative \% of GC} = \frac{G + C}{A + T + G + C} X\ 100$$

Percentage GC content of chromosomal DNA could be determined by subjecting the purified DNA to increasing temperature. Double stranded DNA gets converted to single stranded DNA at high temperature and causes increase in the absorbance at 260 nm. Higher the GC content of a microorganism, higher will be the melting temperature (Tm).

Table. Bae composition of DNA in different bacteria.

Microorganism	% GC
Clostridium tetani	25
Bacillus subtilis	42-43
Escherechia coli	50-51
Rhizobium meliloti	62-66
Azotobacter chroococcum	65-66
Micrococcus luteus	71-76

The degree of similarity or relatedness between two DNA or RNA samples from different bacteria can be determined by measuring the extent of hybridization between denatured and radioactively labelled DNA molecules or between single stranded DNA and RNA species under appropriate conditions (Southern hybridization or Northern blotting).

Table. DNA homologies among certain bacteria as determined by DNA -RNA hybridization.

DNA source	% Relatedness to E. coli
Bacillus subtilis	1
Proteus vulgaris	14
Aerobacter aerogens	45
Shigella dysentreriae	71
Escherichia coli	100

8. Serology

Bacterial species can be identified by specific antigen-antibody reactions. Antigens are defined as proteins or lipopolysaccharide molecules that induce the production of antibodies when injected into animals. The production of antibodies in an animal host could be tested by either the agglutination, immunoprecipitation or immunodiffusion test.

A. ***Agglutination test:*** A drop of bacterial culture (which are naturally coated with antigens) is mixed on the slide with the antiserum/antibody. If clumping occurs, the test bacterium is considered to be the same or closely related to the bacterium used as the antigen. this technique was discovered in 1900 by K. Landsteiner and used in typing of human blood - A, B, AB and O.

B. ***Immunoprecipitation test:*** This serological reaction occurs in two stages. (i) antibodies bind to antigens within seconds or minutes, then (ii) the constant regions of antibodies in these immune complexes bind to each other in a reaction that takes several hours and results in the formation of a visible precipitate. Those concentrations of antigen that result in precipitation of nearly all antigen and specific antibody define the zone of equivanence.

C. ***Immunodiffusion test:*** This test is used to check the similarity of soluble antigens. When a solution of antibodies and a solution of antigens are placed in nearby depressions (wells) o a gel, this bands of precipitin can be visualized in the zone of equivalence. The antiserum diffuses through an air gel towards a source of crude extract of antigens/bacteria and encounters each other. A sharp line of precipitate is formed in the zone of antigen-antibody interactions. Such immunodiffusion techniques are often used to determine whether two antigens share common antigenic determinants. tthe v-shaped white lines of precipitin appear at the zone of optimal concentration between those antigens having mutual reactivity and are termed the reaction of identity. If two completely unrelated antigens are added to the wells, either a single straight band forms between two wells or two separate bands form between two wells, creating an x-shaped pattern, termed a reaction of nonidentity.

Three classes of surface antigens have been extensively studied among various enteric bacteria. these are:

O antigens: Lipopolysaccharides in the outer membranes

K antigens: Capsular polysaccharides

H antigens: Flagellar antigens

9. Ribosomal RNA (rRNA) phylogenetic relationship

In microorganisms, three types of RNAs are found. These are messenger RNA (mRNA), ribosomal RNA (rRNA) and transfer RNA (tRNA).

Ribosomal RNA (rRNA) is found in association with the ribosomes that are involved in protein synthesis. Bacterial ribosomes contain three species of rRNA i.e. 5S, 16S and 23S. Over the last two decades, comparative analysis of 16S rRNA sequences has been used to explore prokaryotic phylogeny. The 16S rRNA is sufficiently large and easy to handle with a high degree of precision (5S rRNA is of small size whereas 23S rRNA is of much larger size). The 16S rRNA isolated from a variety of organisms was digested with ribonuclease T1. The resulting nucleotides are resolved by two-dimensional electrophoresis and sequenced. The comparisons of sequences represent relationship among organisms.

Evolutionary (phylogenetic) relationship: The prokaryotic and eukaryotic cellular forms

(Evolutionary linkages based on ribosomal RNA sequences)

Class: Mollicutes (Tenericutes)

Important group is Mycoplasma. Cell wall is absent. These are obligate chemoheterotrophs with complex nutritional requirements. These are mostly parasites on plants and animals. e.g. Mycoplasma, require sterol whereas Acholeplasma does not require cholesterol.

The mycoplasmas are distinguished by their lack of cell wall, the outer boundary of the cells being the cytoplasmic membrane. As a result, the cells have elasticity and can assume many different shapes ranging from branched filaments. The elasticity allows many of the cells to pass through bacteriological filters. They are susceptible to lysis by osmotic shock caused by sudden dilution of the medium with water. Because of lack of cell wall, mycoplasmas are not inhibited by even high level of penicillin; however, they can be inhibited by antibiotics that affect protein synthesis, such as tetracycline or chloramphenicol.

Mycoplasmas can be cultivated *in vitro* on non-living media as facultative anaerobes or obligate anaerobes. The colonies are embedded in the agar surface and usually have a characteristic fried-egg appearance. Mycoplasmas differ from "L-shape variance" that can develop from other bacteria. Such variants are osmotically fragile, cell-wall-defective forms that from fried-egg colonies resembling those of mycoplasmas. Spiroplasma citri has been isolated from the leaves of citrus plants, where it causes a disease called citrus stubborn disease and from corn plants suffering from corn stunt disease. Four species of *Spiroplasma* are recognized that causes a variety of animal disease such as honey bee spiroplasmosis, suckling mouth cataract disease and lethargy disease of the beetle Melolontha.

Class: Mendosicutes - Archaebacteria

True peptidoglycan is absent in this group. They have pseudomurein including large protein and polysaccharide molecules.

(i) Methanogens: Methane is a prominent end product. They convert fermentation products such as CO_2, H_2, formate and acetate (produced by other anaerobes) to CH_4 and CO_2. They have pseudomurein in their cell wall. They are involved in biogas production. e.g. *methanobacterium*, *Methanococcus* and *Methanosarcina*.

(ii) Halophiles: They prefer to grow in highly saline environments as habitat (salt lakes, brines). e.g. *Halococcus*: which is immotile. *Halobacterium*: rod shaped bacteria with polar flagella. They require high salt concentration. Minimum NaCl concentration required is 2.0-2.5 M (that permits growth). Optimum NaCl concentration required is 4.0-5.0 M. Magnesium (Mg^{++}) requirements is also very high (about 0.025 M). Red carotenoid pigments protect the cell from photochemical damage by high light intensity characteristic of the natural environment. These organisms are aerobic chemoheterotrophs with complex nutritional requirement.

(iii) Thermoacidophiles: They require high temperature and low pH as habitat. e.g. *Sulfolobus* and *Thermoplasma*. They grow fairly at wide temperature 55-85^{0}C and pH 1.0-5.9. The G+C contents are 60-68%. *Thermoplasma* is found in the refuge piles from coal mines and it requires unusual growth factor present in yeast extract. Oxidation of

FeS by chemoautotrophic bacteria acidifies and heat the pile, creating a unfavourable growth environment for *Thermoplasma* (pH 05-4.5; temperature 37-65^{0}C).

Unique properties of prokaryotic organisms

1. Nitrogen fixation: The property to utilize atmospheric nitrogen fro its reduction in plant utilizable form ammonia lies only with prokaryotes. For example, *Rhizobium* fixes N_2 in association with the legumes and *Frankia* (an actinomycete) in symbiosis with actinorhizal plants. *Azotobacter* is a free-living symbiotic N_2-fixer.
2. Anaerobiosis: Some of the bacteria can grow in the absence of oxygen using inorganic compounds (e.g. NO_3 or SO_4) for respiration. These bacteria use fermentative pathways for generating energy rich compounds such as ATP. e.g. *Clostridium, Methanobacterium* and *Methanococcus.*
3. Nature of reserve food material: Polymers of β-hydroxy butyric acid as a reserve material is restricted only to prokaryotes. e.g. *Bacillus*, *E. coli* and *Azotobacter.*
4. Chemoautotrophs: A bacterium having the ability to utilize CO_2 as a source of carbon and generating energy from oxidation-reduction reactions of inorganic compounds is chemoautotrophic. e.g. *Nitrosomonas, Nitrobacter* and *Thiobacillus*.

Measurement of bacterial population

Various methods have been used for qualitative and quantitative measurements of the bacterial population.

Qualitative analysis: Various procedures have been proposed for direct microscopic examination of bac teria in the soil. The extensively used method is Rossi-Cholodny buried slide or contact slide method.

A microscopic slide is buried in the soil and incubated for appropriate period. The slide is removed and the larger soil particles are dislodged carefully. The microbial film adhering to and developing upon the glass surface is stained with phenolic rose Bengal dye and examined microscopically. The Rossi-Cholodny buried slide method allows microorganisms to develop in the physical posture typifying their normal position. It also shows the associative relationships or other microbial interactions with their neighbours taking place in the soil.

In addition to conventional staining techniques and light microscopy, other direct methods of examining soil microorganisms are:

- The use of microcapillaries incubated in contact with soil.
- Examinations of soil suspensions by dark field or phase-contrast microscopy or by the application of fluorescence microscopy with an appropriate agent for the fluorescence technique.
- By viewing the cells in soil suspensions by transmission electron microscopy.

Quantitative analysis

1. ***Haemocytometer method:*** One of the techniques involves the incorporation of weighed amounts of soil in the melted agar and the addition of drops of agar infusion to calibrated haemocytometer. The suspension is stained and then examined microscopically. If the amount of soil, the volume of the agar and the area over which the agar is spread, are known, then the bacterial members can be determined quantitatively.

2. ***Direct microscopic count:*** This technique requires the use of a specialized counting chamber called the Petroff-Hauser Chamber in which the number of cells present in an aliquot of a cell suspension is counted and the total number is determined mathematically. Although it is a rapid method, it has the disadvantage that both living and dead cel,ls are counted. This method is not sensitive to population fewer than 1 million cells.

3. ***Use of electronic cell counter:*** The Coulter Counter is an example of the instrument capable of rapidly counting the number of cells suspended in a conducting fluid that passes through a minute orifice.. Electric current is allowed to pass through the conducting fluid. Cells increase the electrical resistance of the conducting fluid and the resistance is recorded electronically. Thus the number of microorganisms flowing through the orifice is enumerated. This method has the disadvantage that it is unable to distinguish between living and dead cells. This method is also unable to differentiate inert particulate material from cellular materials.

4. ***Chemical methods:*** These methods may be used to measure indirectly the increase in protein concentration or DNA production. In addition, cell mass can be estimated by determining dry weight of the culture. Measurement of certain metabolic parameters and enzymes may also be used to enumerate bacterial population. The amount of oxygen uptake is directly proportional to the increase in the number of vigorously growing aerobic cells.

5. ***Turbidimetric methods:*** Increased turbidity in a given growing culture is another index of growth measured with the help of spectrophotometer. The

amount of transmitted light decreases as the cell population increases. The decrease in radiant energy is converted to electrical energy, which is indicated on a galvanometer. This method is rapid but has limited application because sensitivity is restricted to microbial growth suspension of 10 million cells or greater.

The major disadvantage with all these methods is that the total count includes dead as well as living cells. To determine the viable cell count, a serial dilution agar plating technique is used.

6. *Serial dilution agar plate method:* In this technique, sample of a growing culture or soil suspension is diluted to isolate the bacterium in pure culture as isolated colony and the numbers of colonies appearing on particular medium plates are counted. Dilution requires a mixing of a small amount of sample with a large volume of sterilized water or normal saline, termed as dilution blank. the sample suspension is serially diluted in a series of dilution blanks. This process is known as serial dilution. An appropriate amount 0.1 or 1.0 ml of the different dilutions is added on the sterilized plate and the appropriate medium cooled to 50-60^{0}C is poured over the suspension. The medium is mixed with soil suspension by gentle shaking and incubated for 2-5 days at 28^{0}C in an incubator. Or alternatively, the medium is added in the sterilized plates first and then 0.1 ml of various serially diluted soil samples are spread on the solidified medium plates with the help of pre-sterilized spreader (sterilized with alcohol). The plates are incubated for 2-5 days at 28^{0}C in an incubator. Numbers of colonies growing on different plates are recorded and their population per ml of culture suspension is calculated. Plates containing 30-300 colonies should be considered for counting the number of viable cells. The advantage of this method are that only viable cells are counted by using this method. Secondly, it allows selection of isolated or separated colonies that can be sub-cultured into pure culture which may subsequently be identified.

The disadvantage of using this method are that:

1. Overnight incubation is necessary before colonies develop on the agar surface.
2. More glassware is required in this procedure.
3. Sometimes the errors in dilution making or during plating may give erroneous results.

Calculations

No. of colony forming units = Number of colonies x dilution factor.

The population count is calculated and expresses as colony counts per ml of culture suspension.

7. ***Most probable number (MPN) method:*** Some microorganisms never produce recognizable colonies on agar media. For such type of microorganisms, the most probable number (MPN) technique is used. This is a method of estimating microbial density without direct enumeration.
A known volume of a tenfold soil dilution series is inoculated into the tubes or flasks of nutrient medium broth suitable for the growth of specific microorganisms under study. The growth will commence provided the inoculum contains one or more cells. For example, if the growth is observed in the tubes or flasks containing inoculum for the 10^5 dilution but not from the 10^6 dilution. The population of that microbial type is recorded as between 10^5 to 10^6. Quantification is achieved by inoculating five or more replicate tubes/flasks of medium with each dilution. The number of tubes/flasks showing turbidity or growth is recorded at each dilution. The number of microorganisms present in the given sample is calculated by making use of the most probable number tables.

Limitations in measurement in bacterial population

1. The total bacterial population in any soil is difficult to determine due to heterogeneity in substrate requirement (whether sucrose, mannitol, peptone, tryptone, growth factors), pH of the medium, oxygen tolerance (aerobic, anaerobic or microaerophilic), salt requirements (Mg, Fe, Cl, PO4, Mo etc.) and incubation temperature (15^{0}C, 25^{0}C or 30^{0}C). Thus by incorporation of different substrates, adjusting appropriate pH and incubation conditions of the plates or dilution tubes under aerobic or anaerobic conditions, it is possible to determine the population of only a specific microbial group.

2. Another limitation is the occurrence of compact ,micro-colonies and tight sticking of the bacteria to the organic and inorganic colloids. These compact micro-colonies may not disintegrate when the soil dilutions are shaken.

3. Errors in soil sampling, preparation of sample s and dilution procedures often give variations in viable counts.

4. Soil is not homogeneous in composition and various factors such as moisture, organic matter content, air circulation, pH, soil texture and soil structure within a given location affect bacterial population, thus giving variations in microbial population from one horizon to another soil horizon. For example, the presence of a single rootlet or particle of plant debris may cause microbiological effect of magnitude sufficient to change the counts ten- or a hundred-fold.

A single medium cannot provide adequate nutrients for all the species of bacteria present in soil. Since the growth requirements of many bacterial species/strains are unknown. Thus the observed count represents only a fraction (1.0%) of total microorganisms present in soil ecosystem. Vast differences in population of different bacterial communities have been reported in dissimilar soils. Thus the cell number of different bacteria varies in different soil types. Usually, the populations found on agar media are: 5-60% *Arthrobacter*, 7-67% *Bacillus*, 3-15%, up to 20% *Agrobacterium*, 2-12% *Alcaligens* and 2-10% *Flavobacterium*. The population of *Corynebacterium, Micrococcus, Staphylococcus, Xanthomonas, Mycobacterium* and *Sarcina* has been reported less than 5.0% of the total bacterial colonies.

Problems in morphological and generic groups of bacteria

Due to the vast number and variety of bacteria in soil, bacteria can be subdivided into a number of morphological groupings. Although, difficulty is often encountered in characterizing the soil bacteria on the basis of their morphology, because many of the dominant strains exhibit several shapes in culture depending (i) upon the age of the cells and the (ii) medium used. For example, short rods of *Arthrobacter* change to cocci with age. Similarly, a Gram negative rod shaped bacterium becomes Gram positive as the culture ages. (iii) The shapes and size of many bacteria in soil appear different to those when the organisms are grown in culture media. A high percentage of bacteria in soil are surprisingly small and have diameters of less than 0.3 µm and some smaller than 0.1 µm. Thus a significant portion of the cells in soil would not be seen by light microscope.

Factors affecting bacterial population in the soil

Various environmental factors such as nutrient availability, moisture, soil temperature, pH, soil structure and texture, cultivation pattern, soil depth and seasonal variations greatly affect microbial population in the soil.

Nutrient availability: Available organic and inorganic nutrients affect proliferation of soil bacteria, since most of the soil bacteria re heterotrophs. The organic matter and available nutrients are usually present in the topmost layer (O-horizon) in most of the cultivated soil. Therefore, microbial population and microbial activities are mostly confined to upper 30-40 cm depth of the soil.

Moisture: Moisture affects both aeration as well as biochemical activity of the microorganisms. The most favourable moisture level for aerobic bacterial activities is at 50-75% water holding capacity of the soils. Excess moisture creates anaerobiosis and low moisture levels adversely affect biochemical transformations.

Soil temperature: temperature primarily affects all biological activities in the soil. A bulk of soil bacteria belong to mesophilic category. Besides affecting microbial growth, temperature also affects the different biochemical processes. Warming trends enhance biochemical activity/transformations by bacterial population.

pH: Highly acidic or alkaline conditions tend to inhibit the growth of many common soil bacteria because the optimum pH for growth of most bacterial species is near neutral pH.

Soil structure and texture: Soil structure affects soil porosity. In sandy soils, pores are large but total pore space is less whereas in clay soils, pores are small but the total pore space is more. Soil texture that is determined on the basis of sand, silt and clay content of the soil, affects the total surface area for microbial activity. Soils having greater proportion of clay mineral complexes favour higher microbial activity.

Cultivation pattern: It also has a marked influence on the types and population of soil bacteria. Grasslands usually have more bacteria than barren lands as a result of greater root density and large exudation of utilizable organic compounds from roots along with more availability of decomposed organic matter from plant debris. Bacterial numbers are similarly higher in cultivated land than in the virgin land.

Soil depth: The bacterial population decreases with soil depth. In deeper layers of the soil, the availability of organic and inorganic nutrients becomes lesser. Moreover,. the environment becomes more anoxygenic due to limiting oxygen diffusion rate in lower horizons of the soil.

Seasonal variations: Like precipitation, ploughing and tillage operations also affect soil aeration. thus seasonal variations affect microbial population and biochemical activities.

Significance of bacterial population on soil processes

The biochemical diversity and functioning of bacteria as a group has a major impact on soil processes.

1. Bacteria bring about many of the biochemical transformations in soil and have important implications for the cycling of several key inorganic nutrients. They increase the availability of inorganic nutrients in soils by mineralizing soil organic matter and by solubilizing soil minerals. Thus, they increase soil fertility. e.g. *Pseudomonas, Thiobacillus, Bacillus, Cellulomonas* and *Arthrobacter*. These bacteria, which bring out many of the major biochemical transformations in soil, probably constitute less than 10% of the total bacterial population. Many soil coryneform bacteria including *Arthrobacter* sp. are major catabolizers of heterocyclic compounds and hydrocarbons.

(i) Bacterial species rapidly metabolize the sugars, starches and simple proteins from soil organic matter. But these bacteria decompose slowly other substances such as lignin, waxes and oils in plant residues. these reclacitrant compounds, along with the products of microbial metabolism that resist further decomposition, become part of the stable organic fraction of soils, known as humus. Humus improves soil structure and its ability to retain nutrient ions and water, greatly enhances the soil's capacity to support plant growth. Nutrients are slowly released from this fraction of soil organic matter as specific group of bacteria, notably the actinomycetes, gradually breaks down its complex constituents.

(ii) Nitrogen -fixing bacteria convert atmospheric inert nitrogen to plant utilizable ammonia with the help of enzyme nitrogenase. e.g. *Rhizobium, Azotobacter, Azospirillum. Rhizobium* spp. form N_2-fixing nodules on the roots of leguminous plants. These dinitrogen-fixing bacteria add nitrogen to soil ecosystems in a biological usable form.

(iii) Nitrifying bacteria convert ammonia to nitrate (an inorganic ion), which is a source of utilizable n2 for plants. But nitrate is more acceptable to leaching and denitrification in the soil. Denitrification occurs in the soil when facultative anaerobes use nitrate as the final electron acceptor in anaerobic respiration. Denitrification leads to production of N_2O and NO that diffuses out of the soil in the atmosphere. These gases deplete ozone gas in the atmosphere and thus increase the chances of skin cancer, cataract and inflammatory diseases.

(iv) Phosphate solubilizing or mineralizing bacteria enhance the availability of inorganic phosphorus in soil.

(v) Sulphate reducing bacteria have a similar role in the sulphur cycle. The obligate anaerobes cause reduction of sulphate to hydrogen sulfide as they respire anaerobically with SO_4^{2-} as their final electron acceptor. The H_2S produced by these bacteria usually precipitates as insoluble metal sulfides. Sulphur oxidizing bacteria regenerate SO_4^{2-} by oxidizing reduced sulphur compounds in soil.

2. Bacteria also have important interactions with plants. Soil microbes colonize plant roots and profoundly influence growth and productivity in natural and agricultural ecosystems. Bacteria may increase the availability of inorganic nutrients in soil by mineralizing soil organic matter and solubilizing soil minerals. They may also compete with the plants for available nutrients through the process of immobilization, thereby affecting the growth of plants adversely.

3. Many bacterial species influence plant growth directly by producing hormones or toxins that may stimulate or inhibit root function and morphology. For example, Azotobacter, Pseudomonas and *Azospirillum* produce growth hormones such as indole acetic acid (auxins), cytokinins and ethylene. Some bacteria produce toxins and cause disease in plants. e.g. *Xanthomonas, Erwinia* and *Agrobacterium*.

4. Some plant growth promoting rhizobacteria (PGPR) colonize plant roots as harmless saprophytes and compete with the pathogens for nutrients, water and infection sites whereas other PGPR produce antibiotics, siderophores, hydrolytic enzymes or other secondary metabolites siuch as HCN to kill the pathogens. These bacteria are being investigated as potential biocontrol agents because of their ability to inhibit the growth of plant pathogens. e.g. *Pseudomonas* and *Bacillus*.

5. Bacteria also play an important role in the natural dissipation of xenobiotics in the environment and thus providing the potential for bioremediation. The biotic (microbial) activities are generally more significant because their enzyme systems greatly accelerate degradation rates and may result in complete destruction, or mineralization of organic pollutants. Microbial communities are thus important in the clean up operations of the contaminants, providing the potential for bioremediation. e.g. *Pseudomonas*.

6. Dairy products like curd, yoghurt, cheese etc. are produced with the help of bacteria such as *Lactobacillus* and *Propionibacterium*.
7. Certain products like vitamins (Propionibacterium), amino acids (Corynebacterium), enzymes like protease and amylase (Bacillus) etc. are produced by the bacteria from cheap raw materials.
8. Bacteria are also involved in biogas production from dung of animals and other agricultural wastes. e.g. *Methanobacterium* and *Methanococcus*.
9. Soil may also contain bacteria which are pathogenic to animals and even to humans. For example, Clostridium spp. are common in soil. These spore forming anaerobic organisms such as *Clostridium tetani* and *Clostridium perfringes* can enter wounds and cause the disease tetanus and gas gangrene, respectively. These diseases are lethal and prophylactic immunization is used to prevent tetanus.

Bacillus anthracis, a common soil species, is also pathogenic to humans. It produces spores that cause the disease anthrax. The spores of *Bacillus anthracis* could be used as biological weapon. In the 1940s, the island of Gruinard, off the north coast of Scotland, was infected with spores of *Bacillus anthracis* as part of biological weapon test. The persistence of these spores made of island uninhabitable for human beings until 1987. The spores were eradicated by treating the soil with formaldehyde. Other pathogenic bacteria are *Salmonella typhimurium, Mycobacterium tuberculosis* and *Vibrio cholerae*.

Objective type questions

A. Fill up the blanks with suitable words

i. Breaking of preformed inorganic or organic molecules for energy generation and formation of precursor molecules is called as ____________.

ii. ____________ during the year 1866 suggested a third kingdom known as Protista.

iii. In ______________ Manual, bacteria have been classified in four divisions based upon the type of wall that the cells possess.

iv. Allochthonous microorganisms are also known as ____________.

v. Gram staining was discovered by ____________ in the year ________.

vi. Gram-negative bacteria are those which loosen the colour of crystal violet stain and retain the stain _____________.

vii. _________________ is an example of phosphate solubilising bacteria.

viii. Autochthonous or indigenous microorganisms derive their energy from _________________.

ix. The phospholipid content of Gram negative bacteria accounts for _______% of their cell weight.

x. The optimum temperature range for thermophilic organisms is ranges between __________ ^{0}C.

B. Multiple choice questions

i. Which of the following type of microorganism are unicellular in nature -

a. Bacteria   b. Fungi

c. Protozoa   d. Both a and c

ii. The catabolism and anabolism are associated with -

a. Enzyme specific functions   b. Substrate specific functions

c. Both a and b   d. None

iii. The inner cytoplasmic portion contains which of the following constituents

a. Genetic,   b. Structural

c. Cytoplasmic   d. All the above.

iv. Which of the following are examples of Gram positive bacteria -

a. *Bacillus*   b. *Clostridium*

c. *Streptococcus*   d. All of these

v. Which of the following are examples of Gram negative bacteria -

a) *Rhizobium,*   b) *Pseudomonas*

c) *Azotobacter*   d) All of these

vi. Which of the following bacterial species are rod shaped in nature -

a) *Rhizobium*   b) *Xanthomonas*

c) *Pseudomonas*   d) All the above

vii. Dark brown coloration of humus is due to the presence of ------------------ in soil.

a. *Azospirillum* b. *Herbaspirillum*

c. *Spirochaetes* d. All the above

viii. Which of the following microorganisms directly feed soil bacteria -

a. *Escherichia coli* b. *Salmonella*

c. *Proteus* d. All the above

ix. Which of the following organism could fix atmospheric nitrogen -

a. Actinomycetes b. Azolla

c. Fungi d. Both a and b

x. The most favourable moisture level for aerobic bacterial activities is at 50-75% water holding capacity -

a. 50-75% WHC b. 25-50% WHC

c. 0-50% WHC d. None

Answer keys

A. Fill in the blanks

i. Catabolism, **ii.** Haeckel, **iii**. Bergey's Manual, **iv.** Zymogenous or fermentative, **v**. Cristian Gram in 1884, **vi.** Safranine, **vii.** *Bacillus*, **viii.** Soil organic matter, **ix.** 20-30%, x. 45 to 65^{0}C

B. Multiple choice questions

i. **d** ii. **c** iii. **d** iv. **d** v. **d** vi. **d** vii. **d** viii. **d** ix. **d** x. **a**

C. Descriptive type questions

i. Define Gram staining. Write in brief the differences between Gram positive and Gram negative bacteria.

ii. Classify microorganism according to their temperature requirement with suitable examples for each groups.

iii. Enlist the methods of measurement of bacterial population in soil. Briefly describe any two of them.

iv. Write in brief the problems in morphological and generic grouping of bacteria.

v. Enlist the factors affecting bacterial population in soil. Briefly describe those factors affecting bacterial population in soil.

3

Actinomycetes

Introduction

The term actinomycetes, coined in the late 1800s (actins, meaning 'ray' and myces, meaning 'fungus') is a misnomer. Because actinomycetes have a considerably smaller aerial mycelium than fungi and can produce a large number of asexual spores called conidia, which give the colonies a 'powdery or chalky' look.

The actinomycetes constitute a specialized group of soil bacteria that occur in soils throughout the world. The actinomycetes are prokaryotes confirming cellular structures and functions of true bacteria and their morphological similarity to the fungi (Eukaryotes). Between bacteria and fungi, actinomycetes are a transitional category. On their hyphae, most actinomycetes produce single, pairs, or chains of asexual spores.The diameter of the hyphae varies from 0.5 to 1.0 μm. Hyphae of actinomycetes are highly branched like fungi and their growth in liquid culture results in the formation of pellets or clumps. Some actinomycetes are similar to bacteria in their flagella and cell wall composition. They are sensitive to antibacterial inhibitors.

Characteristics

Actinomycetes are prokaryotic. They form colonies on agar surfaces but the actinomycetes colonies are not gummy or slimy (unlike to that of bacteria). The actinomycetes colonies are powdery in consistency. The colonies firmly adhere to agar surface and crumbles when touched.

Except for *Actinomyces*, the majority of soil actinomycetes have slender, branching filaments that grow into mycelium. Actinomycetes, like fungus, create clumps or pellets in liquid medium. For proliferation, actinomycetes produce single, pairs, or chains of asexual spores (conidia) on their hyphae. When certain genera produce airborne spores, the surface often becomes coloured. The diameter of hyphae, conidia, or individual vegetative cells (0.5 to 1.0 m) is the same as the diameter of the bacterial cell.

Actinomycetes' mycelium lacks chitin and cellulose in its cell walls, and its cell wall composition is comparable to that of bacteria. Actinomycetes have peptidoglycan in their cell walls, which is sensitive to lysozyme, which dissolves the peptidoglycan's polysaccharide backbone.

In terms of shape, physiology, and staining reactions (gram positive response), actinomycetes that do not form mycelial filaments resemble normal bacteria such as Mycobacterium and Corynebacterium. Some motile actinomycetes have flagella that are comparable to bacteria's. Antibacterial inhibitors (antibiotics) are toxic to actinomycetes, but not to antifungal inhibitors.

Actinomycetes population in soil

Actinomycetes are widely distributed in soil, composts and river muds. Most actinomycetes are chemoheterotrophs and their viable counts are sometimes equal to that of bacteria.

***Enumeration*:** For quantitative enumeration, plate count method is used. Both vegetative cells as well as conidia form colonies on medium plates, which usually give a higher count than the actual functional population. Actinomycetes can be counted on the same plates that bacteria are counted on, but specific medium are generally preferred. Because of the slow growth properties of actinomycetes, the incubation period in other media must be slightly longer than that of bacteria.Particularly useful selective media for enumeration of actinomycetes are those containing chitin. This polysaccharide is used by a high percentage of the more common actinomycetes and by a lower percentage of bacteria and fungi. The preferred media used for actinomycetes growth are: Glycerol yeast extract agar medium, Thornton's asparagines mannitol agar media, Kenknight and Munaier's agar media which could be supplemented with chlorotetracycline (aureomycin) to prevent bacterial growth.

In temperate zones, the actinomycetes count in arable soils ranges from 10^5 to 10^7 per gram of soil. In both virgin and developed areas, actinomycetes account for 10-15% of the entire microbial community. Actinomycetes contribute 300-3000 kg wet mass per hectare of soil biomass.In alkaline soils, their population is much richer than other organisms. The population of actinomycetes is more in grass and pasture lands than in uncultivated soils. Actinomycetes are mostly aerobic and therefore, their population is always higher in upper horizon than in the sub-horizons of the soil.

The actinomycetes are poor competitor with bacteria and fungi during initial stages of plant residue composition due to their slow growth rate. The actinomycetes, however, predominate over bacteria and fungi at the second stage of organic matter decomposition because of their ability to decompose complex organic compounds like humic substances of plant and microbial origin, fungal chitin, paraffins, phenols, steroids and pyrimidines, and due to their oligotrophic nature (ability to grow at very low concentration of organic and inorganic nutrients where other heterotrophs fail to grow).

Species of only a few genera dominate the actinomycetes colonies appearing on agar media inoculated with dilute suspensions of soil. *Streptomyces* is numerically dominant, often accounting upto 70-90% of the actinomycetes colonies on most agar media. Rarely, their relative frequency is as little as 5% of the actinomycetes population in some soil.

Nocardia is usually the second most abundant and 10-30% colonies are usually nocardias. *Micromonospora* species are the third most common actinomycetes found on solid media, accounting for less than 10-15% of all actinomycetes. Aerobic species of *Actinomyces* may be frequent inhabitant of some soils. All other genera exist in sparse numbers. The counts of *Microbispora* and *Thermomonospora* were found less than 10^3 per gram in several Japanese soils. The thermophilic genus Thermoactinomyces is common in heating compost heaps.

The musty (mouldy) odour that Streptomyces produce is evocative of freshly turned soil, and it is not unlikely that the rich earthy fragrance of recently ploughed ground is due to the presence of these micro-organisms. The chemical that causes the earthy stench has been identified, and the odorous Streptomyces metabolite that has piqued the public's interest has been dubbed geosmin. Other volatile compounds produced by *Streptomyces*, on the other hand, could be responsible for the distinctive odour.

Taxonomy

Soils contain a large array of distinctly different genera. These can be divided into few families.

Families

1. Streptomycetaceae: Aerial mycelium is widespread, and each chain of spores contains 5 to 50 conidia. Aerial hyphae bearing numerous conidia form distinct chain like arrangements. Divisions of the hyphae form these spores. Conidia have an oval to rod shape when fully grown, and their size and morphology are similar to those of genuine bacteria. *Streptomyces, Microbiospora, and Sporichthya* are some of the common examples.

2. Nocardiaceae: Hyphae typically fragment to yield small rounded or elongate structures. Examples are: *Nocardia* and *Pseudonocardia.*
3. Micromonosporaceae: Hyphae do not disintegrate. Conidia can be found in singles, pairs, or short chains.Examples are: *Micromonospora, Micropolyspora, Thermomonospora, Thermoactinomyces, Actinobifida* etc.
4. Actinoplanaceae: Spores are formed in sporangia. Diameter of the hyphae may be 0.2 to sometimes more than 2.0 µm. Examples are: Streptosporangium, dactylosporangium, Actinoplanes.
5. Dermatophilaceae: Large numbers of spherical motile entities are formed when hyphal fragments split. Example is: *Geodermatophilus.*
6. Frankiaceae: Inhibits root nodules of certain non-leguminous actinorhizal plants. They are unable to grow apart from the plants. Example is: *Frankia.*
7. Actinomycetaceae: There is no genuine mycelium generated. They are anaerobic to facultative anaerobic in most cases. Example is: *Actinomyces.*

Factors affecting actinomycetes population in the soil

The actinomycetes flora is influenced by the surrounding ecosystem in both qualitative and quantitative dimensions. The factors occurring within the ecosystem dictate the dominant stage of the life cycle, the size of the community, its metabolic processes, and the genera and species found in the soil.The primary ecological and environmental factors that generally affect actinomycetes population in the soil include the organic matter status, moisture, soil temperature, pH, soil texture and structure, cultivation pattern, soil depth and seasonal variations.

***Organic matter*:** Available organic and inorganic nutrients affect proliferation of soil actinomycetes, since most of the soil actinomycetes are heterotrophs. Actinomycetes are strongly affected by the presence of accessible carbon, and their numbers are particularly high in organic-rich soils. Crop leftovers, protein derivatives, and animal manures are examples of organic substances that increase the abundance of actinomycetes. With crop residue turnover, the population can sometimes exceed 10^8 per gram, especially in situations with high temperatures.

When organic matter is added, the bacterial and fungal flora normally thrives at first, especially if nitrogen is abundant. The actinomycetes' response is frequently delayed until later phases of the degradation of plant wastes. This suggests that certain bacteria and fungi are the initial agents of decomposition due to their higher growth rates and biochemical versatility, whereas actinomycetes only appear after the readily available compounds (such as cellulose and hemicellulose) have been metabolised and competitive stress has been reduced.

***pH*:** Low pH is toxic to these microbes. Below pH 5.0, most Streptomyces strains cease to multiply or have little activity. Actinomycetes are rarely more than 1.0 percent of the total viable count in severely acidic environment. The sensitivity of actinomycetes to low pH in the range of 5.0 has practical applications in the treatment of some plant diseases caused by Streptomyces, in which soil acidification is employed to restrict pathogen growth. e.g. Potato scab disease caused by *Streptomyces scabies.*

Actinomycetes have been shown to be suppressed by the continual use of ammoniacal fertilisers. The action of microorganisms in the soil converts ammonium to nitric acid, and the resulting drop in pH causes unfavourable growth circumstances. Liming the soil is useful because neutral or alkaline conditions favour vegetative development, with the population being most plentiful in soils with pH 6.5 to 8.0.

***Moisture*:** Another important environmental variable is moisture. It has an impact on the microorganism's aeration as well as its metabolic activities. The best moisture level for aerobic bacterial activity is 50-75 percent of the soil's water holding capacity (WHC). These microorganisms only appear in extreme cases of waterlogging or when moisture levels are above the microbiological optimum (i.e. 85 to 100 percent of WHC). Anaerobiosis is caused by too much wetness. When free oxygen is scarce, all common actinomycetes are unable to proliferate and spread due to their aerobic metabolism.

Actinomycetes, on the other hand, are not as affected by semi-dry environments as bacteria are. Low moisture levels favour the vegetative development and conidia production of filamentous actinomycetes. As a result, as soils dry up, the quantity of actinomycetes remains high but the relative incidence of bacteria decreases due to bacteria's lack of tolerance for arid environments. In some real deserts, actinomycetes predominate, however the moisture effect of this type is linked to conidia persistence. These findings show that actinomycetes spores are more resistant to desiccation and last longer than those of other taxonomic groups.

***Soil temperature*:** All biological activities in the soil are primarily influenced by temperature. Temperature influences a variety of metabolic processes in addition to microbial growth. Mesophilic actinomycetes grow very slowly at 50°C and almost none at 39°C, according to hyphal density on glass slides embedded in the soil. Temperatures between 50 and 270 degrees Celsius promote the growth of actinomycetes, with the ideal range being between 28and 37degrees Celsius.

Thermophilic cultures are also prevalent, with facultative thermophiles growing at temperatures ranging from 55°C to 65°C as well as 30°C. Obligate thermophiles, on the other hand, do not proliferate at lower temperatures. The existence of thermophilic actinomycetes has been found in soil, manure, heating hay, and compost piles, as well as in soils that have never been warm. The population of thermophilic actinomycetes increases dramatically when manure piles are heated. At 50°C to 65°C, for example, up to 10^{10} population of these actinomycetes can be found per gram of soil. Thermophilic actinomycetes strains are classified in the genus *Thermoactinomyces*, but *Streptomyces* species frequently predominate.

***Seasonal variations*:** Soil aeration is affected by rainfall, ploughing, and tillage operations. Seasonal changes have an impact on the microbial population and metabolic processes. The highest counts of actinomycetes are usually found in the spring and autumn. The return of plant remains to the soil is usually ascribed to the increase in the fall time of the year. Frost killing is thought to be the cause of shortage in the temperate zone during the winter. A comparable reduction occurs during dry seasons of the year, although because of their strong endurance to desiccation, the relative fraction of actinomycetes is often highest during these months.

***Soil depth*:** The filamentous actinomycetes can be found in the A-horizon and at great depths below the surface; however the cell density assessed by plating techniques decreases as the profile deepens.The organic matter and available nutrients are usually present in the topmost layer (O-horizon) on most of the cultivated soil. In deeper layers of the soil, the availability of organic and inorganic nutrients becomes lesser. Moreover, the environment becomes more anoxygenic due to limiting oxygen diffusion rates in lower horizons of the soil. Therefore, microbial population and microbial activities are mostly confined to upper 30-40cm depth of the soil.

The percentage of actinomycetes in the total microbial population increases with depth in some soils, and they make up a significant portion of the subsurface community. This strange behaviour could be due to the downward movement of conidia in water or a difference in the effects of O_2 and CO_2 on bacteria and actinomycetes. Counts may vary from 10^3 to 10^5 per gram of soil even in the C-horizon.

***Soil structure and texture*:** Soil porosity is influenced by soil structure. Pores in sandy soils are large, but total pore space is modest, whereas pores in clay soils are small, but total pore space is high. The total surface area for microbial activity is affected by soil texture, which is defined by the amount of sand, silt, and clay in the soil.Soils having greater proportion of clay mineral complexes favour higher microbial activity.

Cultivation pattern: It also has a significant impact on the actinomycetes' kinds and populations. Because of larger root density and substantial exudation of utilisable organic compounds from roots, as well as more availability of degraded organic materials from plant waste, grasslands frequently have more actinomycetes than barren regions. Bacterial counts on cultivated land are similar to those in virgin land.

Significance and impact of actinomycetes population on soil processes

Actinomycetes have a significant impact on soil because of their biochemical variety.

1. In comparison to most fungi and bacteria, actinomycetes are poor competitors. Their low competitiveness could explain their paucity in the early stages of plant waste degradation. When nutrients become scarce and the pressure from more effective competitors such as bacteria and fungi diminishes, actinomycetes become more prominent.

2. Actinomycetes are heterotrophic feeders; therefore the availability of organic substrates influences their occurrence. Actinomycetes can use a wide range of carbon sources, from simple organic acids and sugars to more complex compounds including polysaccharides, lipids, proteins, and aliphatic hydrocarbons. Many actinomycetes species breakdown cellulose, however the rate of decomposition is always slow.

   Chitin hydrolysis is especially characteristic of frequently encountered species of *Streptomyces* and the addition of this polysaccharide to the soil serve as a means of marked stimulation of *Streptomyces* proliferation. The ubiquitous ability of Streptomyces and other actinomycetes to use chitin as a carbon source has led to the usage of chitin-containing media for isolating members of this group.

   Proteins, lipids, starch, and inulin can all be degraded by actinomycetes. *Micromonospora* strains breakdown chitin, cellulose, glucosides, and hemicelluloses, whereas *Nocardia* species have well-documented metabolism of paraffins, phenols, steroids, and pyrimidines. Certain actinomycetes isolates thrive in low-carbon environments, and are classified as oligocarbophilic (oligotrophic) bacteria.

3. On the roots of actinorhizal plants, Frankia develops N_2-fixing nodules. Frankia species that nodulate on non-legumes like *Alnus* (alders, e.g. red and black alders) fix about 40-300 kg of N_2 ha^{-1} yr^{-1}, while *Casuarina* (Australian pine) fixes about 58 kg of N_2 ha^{-1} yr^{-1}.

4. Many strains in the order Actinomycetales have the capacity to synthesize toxic metabolites. Antibiotics are antimicrobial compounds produced by around three-quarters of Streptomycetes isolates. The medically important antibiotics derived from species of *Streptomycetes* include streptomycin, neomycin, erythromycin and tetracycline etc. Antibiotics generated in culture can impede the growth of bacteria, yeasts, and fungus from a wide range of taxonomic groupings.The percentage of actinomycetes producing antibiotics varies with the soil and season of the year. Despite their high industrial and medicinal utility, the importance of antibiotics in natural ecosystems such as soil has yet to be determined. Because the formation of antibiotics by actinomycetes actually living in the soil is hard to demonstrate, and if at all occurs in soil to a slight extent owing to unsuitable nutritional conditions. Further, any antibiotic effect would be limited to the vicinity of the colony responsible because antibiotics are inactivated by adsorption onto the surfaces of clay minerals and are destroyed by non-biological and biological reactions.

5. In addition to producing antibiotic compounds that lyse bacteria, several *Streptomycetes* species release extracellular enzymes. The presence of these enzymes may play a role in the microbial balance in soil.

6. Because microscopic analysis reveals that few actinomycetes are in the mycelial stage, and the high plate counts are mostly the product of conidial persistence, actinomycetes have a lower biochemical role in soil nutrient transformation activities than bacteria and fungi. Nonetheless, there is evidence that these bacteria are involved in the processes listed below.

   Plant and animal cells decompose certain resistant compounds.These actinomycetes fare poorly in competition with bacteria and fungi when simple carbohydrates are present in the added carbonaceous materials. Only when resistant compounds are left in the added plant residues do they usually become effective competitors.

   Humus is made by converting plant remnants and leaf litter into compounds that are native to the soil organic fraction. Many actinomycetes strains can create the many complex compounds that are thought to be crucial in mineral soil humus fractions. High-temperature transformations, especially in the rotting and heaping of green manures, hay, compost piles, and animal manures. The thermophilic actinomycetes may be a dominating group under these conditions. Because of actinomycetes growth, the surface of the compost pile might sometimes take on the white or grey colour associated with the actinomycetes group. *Thermoactinomyces*, certain *Streptomyces* species, and some spore-forming bacteria have a competitive advantage under these conditions.

7. Actinomycetes are responsible for a number of plant diseases that are spread through the soil. Potato scab is caused by *Streptomyces scabies*, and sweet potato pox is caused by *Streptomyces ipomoeae.*

   Some of the actinomycetes cause infections to humans and animals. For example, *Nocardia asteroids* and *Nocardiaostitidis-cavirum*. In people who have developed hypersensation to antigens in the dust, inhaling dust from mouldy hay can cause an allergic reaction known as 'Farmer's lung'. Spores of thermophilic actinomycetes notably *Thermoactinomyces vulgaris* and *Micropolyspora faeni* are the principal sources of this antigen. The spores of these actinomycetes are small and they can penetrate deeply into the respiratory tract, in the peripheral parts of which the allergic symptoms develop.

8. Actinomycetes also play a role in microbial antagonism and microbial community composition regulation. The ability of many actinomycetes to secrete antibiotics or create enzymes that cause fungus and bacteria to lyse could explain their importance in the environment.Thus several species of actinomycetes probably play a major role in maintaining microbiological equilibrium in the soil.

9. The addition of chitin to the soil, which promotes actinomycetes hyphal development, can sometimes result in a significant reduction in the growth of disease-causing fungi. As a result, several actinomycetes play a role in biological disease management in higher plants.

Objective type questions

A. Fill up the blanks with suitable words

i. ______________________ is an example of antibiotic produced by actinomycetes.

ii. In temperate zones, the actinomycetes count in arable soils ranges from __________ to ________ per gram of soil.

iii. Mesophilic actinomycetes generally grows in the temperature ranges between _______ to _______ ^{0}C.

iv. _______________________ is an example of actinomycetes species that could fix atmospheric nitrogen.

v. An example of cellulose decomposing algae is __________________.

vi. _________ and ___________ are non-leguminous plants that could fix atmospheric nitrogen.

vii. Potatoscabiscausedbytheactinomycetesspecies__________________.

B. Multiple choice questions

i. In temperate zones, the actinomycetes count in arable soils ranges from -------------- of soil.

a. 10^5 to 10^7 per gram
b. 10^6 to 10^8 per gram
c. 10^7 to 10^9 per gram
d. None

ii. Which of the following media used for actinomycetes growth

a. Glycerol yeast extract agar medium,
b. Thornton's asparagines mannitol agar media
c. Both a & b
d. None of these

iii. Which of the following actinomycetes species is dominant in soil

a. *Streptomyces*
b. Nocardia
c. Micromonospora
d. None

iv. Actinomycetes generally prefers which of the following soil reaction

a. Acidic
b. Neutral
c. Slightly alkaline
d. Saline

v. The best moisture level for aerobic bacterial activity is

a) 50-75 % of WHC
b) 60-80 % of WHC
c) 80-100 % of WHC
d) None

vi. Which of the following non-legume species could fix atmospheric N_2 in association with actinomycetes

a Alnus b Myrica

c Casuarina d All of these

vii. Which of the following soil horizon has highest actinomycetes population

a. O-horizon b. A-horizon

c. B-horizon d. C-horizon

Answer keys

A. Fill in the blanks

i. Streptomycin or Tretracycline, **ii.** 10^5 to 10^7 per gram, **iii.** 35-40^{0}C, iv. *Frankia,* **v.** *Micromonospora*, **vi.** *Alnus and Myrica,* **vii.** *Streptomyces scabies*.

B. Multiple choice questions

i. **a** ii. **c** iii. **a** iv. **c** v. **a** vi. **d** vii. **b**

C. Descriptive type questions

i. Classify actinomycetes taxonomically with suitable examples.

ii. Write in brief the factors affecting actinomycetes population in the soil

iii. Write in brief the methodology involved in actinomycetes population in soil

iv. Write in brief the characteristics of actinomycetes.

4

Fungi

Fungi are a diverse group of multicellular eukaryotic organisms with wide variation in vegetative and reproductive morphologies and diverse life cycles. They are more abundant, on a mass basis, in soils than any other group of microorganisms; their biomass ranges from 500 to 5000 wet kg ha^{-1} (Metting, 1993). Fungi inhabit almost any niche containing organic substrates, so that they are active participants in ecosystems as -

- degraders of organic matter
- agents of disease as well as beneficial symbionts - lichen (algae and fungi), mycorrhizal association
- agents of soil aggregation, and
- an important food source for humans and many other organisms. Humans depend considerably on fungi fro metabolic by-products that are used in food additives (citric acid, amino acids, single cell proteins) and medicine (antibiotics).

In many cases, they are vital component of ecosystem function and vitality. At least 70,000 species have been described but at least 10 times that number are estimated to exist worldwide (Hawksworth, 1991).

The branch of science that deals with the study of fungi is called Mycology (Greek words, mykos - mushroom and logos - discourse), and the branch that deals with the study of fungal diseases is called fungal pathology.

These fungi vary from gliding, bacterium-engulfing slime molds to wood decaying, mycorrhizal and root pathogenic basidiomycetes with differentiated trophic mycelia and foraging root like rhizomorphs. Soil fungi range from the microscopic, in which the whole organism consists of a single cell with dry weight of less than $1x10^{-12}$ g, to the immense in which filaments of a single clonal microorganism can occur over an area of 15 hectares and the individual has been estimated to have a weight of $1x10^{7}$ g (Smith et al., 1992).

Characteristics

Fungi form a large part of the total microbial biomass in most of the well aerated and cultivated soils. Fungi possess filamentous mycelium network composed of hyphal strands. The hyphae are broad (5 μm in diameter), may be uni-, bi- or multinucleate, septate or non-septate. The propagation is either by asexual spores (conidia) formed by mitotic division or by sexual spores involving gametic fusion followed by meiotic division. The size, shape and colour of conidia or spores, morphology of hyphae and physiological characteristics, form the basis of taxonomic classification of fungi. All fungi are aerobic chemoheterotrophs and therefore can derive the energy and carbon from organic matter present in the soil.

Lichens are stable symbiotic associations between fungi and green algae or cyanobacteria, in which the fungal partner (mycobiont) form a characteristic structure (thallus) that encloses and protects the algae or cyanobacterium (photobiont). Lichens play vital roles in soil biogenesis and stabilization. Mycorrhizal fungi provide plants with phosphorus nutrients and protect from drought stress and root pathogens. Approx. fourth-fifth of the world's vascular plants are known to form arbuscular (endo) mycorrhizae and many of remainder form ectomycorrhizae and only a minority of plants are non-mycorrhizal. The vegetative body of a fungus is called a thallus (not differentiated into root, stem and leaves), and it generally exists as one of four forms:

(i) **Non-mycelial thalli - Plasmodium:** Acellular, naked, multinucleated and amoeboidal mass of protoplasm is known as Plasmodium. The entire thallus gets transformed into several reproductive nits (holocarpic polycentric). e.g. *Physarum, Stemonitis, Trichia* etc.

(ii) **Chytrid cells:** Solitary globose cells with or without specialized rootlike filaments called rhizoids. They are unique to the organisms of the phylum chytridiomycota in the kingdom fungi and the phylum hypochytridiomycota in the kingdom Stramenopila.

(iii) **Yeast cells:** Spherical to ovoid cells are formed by fungi in the phyla Ascomycetes and Basidiomycetes. They divide by budding or fission. Some fungi are dimorphic and can change from a mycelium to a yeast form under conditions (where penetration of a substrate is not needed to obtain nutrients) such as in aqueous environments or insect cavities.

(iv) **Mycelial:** A filamentous network of hyphae that branch and grow only by apical (tip) extension. The vegetative growth form is the most common of organisms in the kingdom fungi and in soil fungi especially. In amycelium, the growth is open ended (closed growth-animal) for as long as a nutrient source is available. The result is that size is highly variable, ranging from a pinpoint to amass covering 15 hectares of land. Mycelial growth measurements show approx. 10-100 metre long hyphae per gram of soil, but values of upto 500m and sometimes in excess of 1000m have also been obtained. On an agar medium in petridishes or on other flat surfaces, a mycelium forms circular colony mats. In a liquid broth medium (where no surface boundaries confine growth), a growing mycelium resembles spherical cotton balls.

Taxonomy

Taxonomically soil fungi are classified into the following major groups.

A. Sexual reproduction (perfect stage) not reported. e.g. Fungi imperfecti or Deuteromycetes.

B. Sexual reproduction (perfect stage) known.

(i) Mycelium aseptate - oomycota (lower fungi)

Class - Phycomycetes

Sub-class - a. Oomycetes, b. Zygomycetes.

(ii) Mycelium septate - Eumycota (higher fungi)

Class - Ascomycetes

Class - Basidiomycetes

1. Phycomycetes (Oomycetes)

- Thallus is mycelial. The hyphae are coenocytic (i.e. aseptate and multinucleate). Asexual reproduction occurs by formation of spores produced inside the sac-like sporangia.
- Some phycomycetes occur on the surface of decaying plant or animal material.
- Some phycomycetes have alternate haploid and diplod generations, and produce diploid mitospores in mitosporangia and haploid meiospores in meiosporangia.
- The sexual reproduction may be isogamous (similar size gamete) or oogamous (male gamete small and flagellated whereas female gamete is large and non-flagellated).
- The mode of sexual fusion is gametangial contact type. Plasmogamy is immediately followed by karyogamy and meiosis.

- Sexual reproduction leads to the formation of resting oospores. That is why phycomycetes is also called oomycetes.

Examples: *Phytophthora infestans* (Oomycetes) causes late blight of potato. *Albugo candida* causes white rust of crucifers. Phythium debaryanum causes damping-off disease of seedlings of tobacco, tomato, mustard and chillies.

Zygomycetes

- Thallus is mycelial and the hyphae are aseptate and coenocytic.
- Cell wall is made of chitin or fungal cellulose.
- Zygomycetes are characterized by absence of any motile (flagellated) body.
- Asexual reproduction occurs by the formation of thin-walled, non-motile sporangiospores, formed by cleavage inside the sporangia. The sporangia are borne at the tips of special hyphae, called sporangiophores.
- Sexual reproduction occurs by gametangial copulation, called conjugation. That is why zygomycetes are also called conjugation fungi.
- Sexual reproduction results in formation of diploid zygospores. during germination, the zygospore gives rise to a promycelium that bears a germ sporangium at its tip. The germ sporangium produces meiospores, called germ spores.

Examples: Mucor pusillus causes bronchomycosis disease in man. Rhozopus stolonifer causes soft rot of sweet potatoes and apples during storage.

Eumycota: Septate Mycelium

2. Ascomycetes

- Ascomycetes are a long group with over 30,000 species. Most of them are terrestrial and occur as saprotrophs or as parasites.
- The yeasts forms are unicellular and among the simplest of sac fungi.
- The ascomycetes are characterized by complete absence of motile structures in the life cycle.
- Majority of ascomycetes reproduce asexually by the formation of conidia or conidiospores. These are non-motile mitospores produced exogenously (mostly in chains) from the tips and sides of hyphae. Conidia are borne on special hyphae, called conidiophores.

- Sexual fusion occurs by gametic copulation (e.g. yeasts)., gametangial contact (*Pyronema*), spermatization (e.g. *Ascobolus carbonarius*) and somatogamy (*Peziza*)
- Sexual reproduction takes place in three stages - plasmogamy, karyogamy and meiosis. Plasmogamy (having two nuclei - dikaryon) is not immediately followed by karyogamy. The intervening phase between plasmogamy and karyogamy is called dikaryophase and cells of this phase are dikaryotic cells.
- Some specialized dikaryotic cells of dikaryophase act as ascus mother cells in which the two nuclei of opposite mating types fuse (karyogamy) to form a diploid nucleus. the resulting diploid nucleus divides by meiosis and then by mitosis to form 8 (sometimes only 4) haploid nuclei which result in formation of 8 haploid ascospores. These sexually produced ascospores (of two mating types) remail inside the sac-like structure called ascus.

Examples

Aspergillus (black smoky mould): *A. fumigatus* causes pulmonary aspergillosis and ear infections.

A. flavus contaminates dry food stuffs such as nuts and groundnuts and produces aflatoxins.

A. oryzae used in preparation of alcoholic beverage 'sake' in Japan.

A. niger used in citric acid and gluconic acid fermentation.

Penicillium chrysogenum (green mould) uses for antibiotic production.

Neurospora (pink bread mould) - Beadle and Tatum - One gene one enzyme theory.

Claviceps purpurea causes ergot of rye. The disease also appears in wheat, barley and oat. Eating of infected cereals causes ergotism. The disease results in convulsions, paralysis, hellucination (a delusion) and finally death to man. Certain ergot alkaloids are used in medicine. The drug ergotine (a mixture of alkaloids) is widely used for the cure of uterine troubles, homorrhages after child birth, migraine and enlarged prostrate glands.

Yeasts: Ascomycetes

- A group of soil fungi commonly referred as yeast occur in soil. These are normally unicellular organisms and reproduce by budding or fission.

- The genera of soil yeasts most frequently isolated are *Candida, Cryptococcus, Debaromyces, Hansenula, Lipomyces, Pullularia, Rhodotorula, Saccharomyces, Schizosaccharomyces, Sporolobomyces, Torula, Torulopsis, Trichosporon* and *Zygosaccharomyes*.
- Yeasts have been found significant numbers in soils of Antarctica, grasslands, cultivated fields and forests. Sometimes, they are particularly numerous on the roots of certain plants.
- Generally populations of approx. 10^3 per gram soil are observed in temperate climates.

3. Basidiomycetes

- The mycelium of basidiomycetes is composed of filamentous, branched and septate hyphae.
- Some of the reproductive cells of secondary mycelium (consisting of dikaryotic or binucleate cells) become club shaped basidia (basidium, singular). These basidia may be septate (phragmobasidia) or aseptate (holobasidia).
- Each basidium produces four (4) haploid basidiospores exogenously at the tip of tubular outgrowths, called sterigmata.
- Motile structures are completely absent in the life cycle.
- Some higher forms produce fruiting bodies called basidiocarps (e.g. mushrooms, bracket fungi etc.). The body of basidiocarp comprises of complex secondary mycelium called tertiary mycelium.
- The basidiomycetes are highly specific in nutritional requirement and difficult to grow on normal agar media.
- They play important role in initial rotting of wood tissues, in lignin decomposition and in ectomycorrhizal association.

Examples

Agaricus campestris is common edible mushroom.

Puccinia graminis tritici - causes black stem rust of wheat.

Ustilago tritici and *U. hordei* causes loose smut of wheat and covered smut of barley, respectively.

Amanita phalloides (death cup) - deadly poisonous.

Armillaria mellea - honey mushroom - fruiting bodies edible.

4. Deuteromycetes: The fungi imperfecti

- They are saprotrophs in soil and on decaying organic matter. Most of them become parasites and cause serious diseases in plants, animals and human beings.
- This class includes those fungi in which sexual stage is not known. It includes 1680 genera and 17000 species.
- The mycelium consists of well-developed, branched and septate hyphae. Cells are usually multinucleate.
- Reproduction occurs only by asexual methods by the formation of conidia. The conidia are either directly produced on conidiophores or in special reproductive bodies called sporodochia, pycnidia and synnemata. Sometimes asexual reproduction also occurs by the formation of oidia and chlamydospores.

Examples

Fusarium oxysporum causes wilt disease in pigeon pea, cotton, tomato, banana etc.

Alternaria solani causes disease (blight) in potato and tomato plants.

Colletotrichum falcatum causes red rot of sugarcane.

Cercospora arachidicola and *C. personata* causes tikka disease in groundnut.

Other phytopathogenic fungi are: *Curvularia, Cylinderocarpon, Helminthosporium, Geotrichum, Humicola, Metarrhizum, Paecilomyces, Rhizoctonia* (root rot), *Trichoderma, Trichothecium* and *Verticillium* (wilt).

Function and activity of fungi

Antibiotic production

Fungi also produce antibiotics. The production of antibiotics by fungal species in ascomycete genera such as *Penicillium* and *Fusarium* provides an ecological advantage in sustaining occupation of a nutrient source. The cereal vascular pathogen *Cephalosporium gramineum* produces a broad spectrum cephalosporin antibiotic that inhibits other organisms from colonizing infested wheat straw when soil pH is less than 6.5. The antibiotic appears to play more of a role in survival than disease development, since its presence or absence has no effect on disease-causing ability.

Antibiotic	Fungi
Penicillin	*Penicillium notatum, P. chrysogenum*
Citrinin	*Penicillium crinium*
Griseofulvin	*Penicillium grisafulvam*
Cephalosporin	*Cephalosporium acremonium, C. gramineum*
Clavicin	*Aspergillus clavatus*
Jawaharene	*Aspergillus sp.*

Jawaherene antibiotic known to be effective against leukemia and many viruses was isolated from a fungus found on rotten potato by Dr. D. K. Roy (1963).

Food spoilage

The detrimental activities of some fungi include spoilage of food by rots, mildews and rusts found on fruit, vegetables and grains. Some *Aspergillus* species are capable of producing toxins (aflatoxins and mycotoxins).

Industrial importance

The fermentative fungi are also of industrial importance.

Wine and beer	*Saccharomyces cerevisiae*
Bread and bakery products	*Saccharomyces*
Amylases	*Aspergillus*
Single cell protein	*Aspergillus, Fusarium*
Gluconic acid, citric acid	*Aspergillus niger*

Mushrooms

Many of the mushrooms produced by the fungi i.e. *Agaricus campestris* and *Agaricus bisporus* are prized for their excellent flavour and large protein content (21-30%).

Some fungi (mushrooms) are notorious for their deadly poison e.g. *Amanita phalloides* (also called death cap) which is highly poisonous.

Lentinus edodes is the edible mushroom of Japan and China.

Morchella esculenta (Guchi) and other mushrooms are rich in proteins, vitamins, carbohydrates, minerals and amino acids. Guchi is sold as costly as Rs. 2000/kg.

Food products

Products	Substrate	Organisms used in fermentation
Idli	Rice	*Torulopsis, Candida*
Jalebies	Moong pulse	*Saccharomyces bayanus*
Kanji	Urd pulse	*Hansenulla anamola*
Shou or soya sauce	Soybean	*Aspergillus oryzae, Saccharomyces rouxii*
Lao Chao	Rice	*Rhizopus chinensis, Endomycopsis sp.*
Ontzom	Peanuts	*Neurospora sitophilia*

Pigments

Catenarin	Red coloured pigment	*Helminthosporium*
Spinulosin	Blue purple pigment	*Penicillium spinulosum*
	Maroon coloured pigment	*Aspergillus fumigatus*

Gibberelin hormone: Produced by *Gibberella fugikuroi* act as plant growth hormone.

Spoilage of goods

Many commercial and household products are attacked and destroyed by fungi, e.g.

Cotton and rayon are affected by fungi *Rhizopus, Mucor, Penicillium* and *Aspergillus*.

Woolen garments are destroyed by *Fusarium, Alternaria* and *penicillium*.

Books are attacked by *Aspergillus* and *Alternaria*.

Jute products are attacked by *Chaetomium globosum*.

Leather products are destroyed due to attack of *Aspergillus niger*.

Timber woods are destroyed by *Polyporus schweinitzii*.

Organic matter decomposition

The major function of filamentous soil fungi is to degrade organic matter as the fungi has the ability to utilize a vast variety of carbon sources including simple sugars, disaccharides, organic acids, starch, pectins, fats, lignins, bacterial polysaccharides, proteins and nucleic acids. Soil fungi such as *Trichoderma, Aspergillus, penicillium* and *Fusarium* are important in the initial extracellular depolymerization of cellulose.

Several fungal species synthesize humic substances in soil while degrading organic matter and thus help in the maintenance of soil organic matter e.g. *Alternaria, Aspergillus, Cladosporium, Humicola* and *Metarrhizum*.

Plant pathogens

Several species of genera *Fusarium, Phytophthora, Pythium, Rhizoctonia, Sclerotium, Verticillium* are active plant pathogens, which could survive in the soil in absence of host plant. *Gaeumannomyces graminis*, the causative agent of take-all disease of wheat, over winters in soil on debris from the previous crop ready to infect the following crop.

The specialized vascular wilt fungi (Fusarium sp. and Verticillium sp.) can only attack juvenile roots but once inside the root, they escape into vascular system. Some fungal species of phyllum oomycota such as *Phytophthora infestans* (causing late blight of potatoes) and *Plasmophora viticola* (causing downy mildew of grapes) produce devastating plant diseases.

Some Basidiomycetous fungi cause tree root and crown rots or cause destructive flower and foliar diseases such as smuts and rusts.

- *Puccinia graminis tritici* - black stem rust of wheat
- *Ustilago tritici* - loose smut of wheat

Some animal and human pathogenic fungi are also encountered in soil.

- *Histoplasma capsulatum, Coccidiodes immitis* and *Cryptococcus neoformans*.
- *Mucor pusillus* causes bronchomycosis disease in man.
- *Aspergillus fumigatus* causes pulmonary aspergillosis and ear infections.

Spores and conidia produced by fungi *Aspergillus, Cladosporium, Fusarium*, rusts, smuts, etc. cause various types of allergies in humans. Some people develop asthma due to fungal allergy.

Lichens

The association of specific algae or cyanobacteria with fungi results in symbiotic association, so called lichens. In this symbiotic association, the algae as the photobiont receives physical protection and nutrients whereas the fungi (mycobiont) obtains carbon and fixed nitrogen from the phycobiont.

Lichens having cyanobacteria as phycobiont possess N_2-fixing activity e.g. Nostoc, Gleocapsa, Calothrix. Examples of N_2-fixing lichens include:

Collema, Nephroma, Peltigera, Solorina, Lobaria, Placynthium, Lichina, Lepogium, Epheba and *Dendriscolaulon*.

These lichens are obviously of ecological importance because of their N_2-fixing ability and their capability to establish in harsh environments such as nutrient poor soils and even on the surface of rocks, where they cause weathering of rocks under low nutrient carbon and nitrogen conditions. Lichens play vital roles in soil biogenesis and soil stabilization.

Mycorrhizal association

Some fungi form mycorrhizal association with higher plants and help in the nutrient uptake of phosphorus and survival under drought conditions. The mycorrhizal fungi form both ectotrophic as well as endotrophic associations. About 4000 fungal species belonging primarily to Basidiomycota and fewer to the Ascomycota are known to form ectomycorrhiza. The ectotrophic fungi are *Boletus, Lactarius, Amanita, Pisolythus* and *Elaphomyces*.

Endotrophic mycorrhizal fungi (VAM fungi) include *Glomus, Gigaspora, Acaulospora, Enterophospora, Scutellospora, Endogene* and *Sclerocystis*.

Phosphate solubilisation

Several heterotrophic fungi secrete acidic substance during their growth which help in solubilisation of insoluble bound phosphates in soil. Members of the genera *Aspergillus, Trichoderma, Paecillomyces, Penicillium* and *Curvularia* have been found to solubilise insoluble bound phosphate compounds in soil.

Composting

It involves microbiological conversion of organic wastes (i.e. crop residues, city garbage, sewage sludge, industrial and animal wastes) into humus and mineral nutrients (organic fertilizer) by the combined activity of bacteria, actinomycetes, fungi and protozoa. Thus, this process serve as a mean of environmentally acceptable waste disposal and produces organic fertilizer required for sustainable soil and crop productivity.

The inocula of selected species of fungi such as *Aspergillus, Trichorus, Trichoderma, Paecillomyces* and *Penicillium* have been found to play appreciable role in promoting composting process.

Bioremediation technology

The white rot fungus *Phanerochaete chrysosporium* secretes extracellular enzymes capable of degrading polychlorinated biphenyls (PCB), polycyclic aromatics such as benzophyrine and DDT.

Biopesticides

A large number of fungi are mycoparasites to insect-pests, particularly *Beauveria bassiana, Metarrhizium anisopliae*, *Trichoderma* and *Verticillium lecani* which are used on commercial scale in USA and European countries. These fungi are ideal for use as biopesticides because these can be easily mass-produced at a reasonable cost.

Various species of *Trichoderma* are very effective in suppression of diseases like chickpea wilt, cotton root rot and damping off disease of various crops. Some of the fungi are nematophagous whereas others are mycoparasites to insect-pests. These fungi are being exploited for use as biopesticides. e.g. *Trichoderma.*

Occurrence

The most common method of determining fungal population is the dilution plate count in potato dextrose agar (PDA) medium having low pH (5.5 or below) and added with rose bengal. Streptomycin, aureomycin, penicillin or novobiocin is added to inhibit the growth of bacteria and actinomycetes. This technique has limitations as fungi are profusely sporulating and each spore as well as each fragment of the mycelium has the ability to develop a colony. Therefore, the colony counts hardly speak of the fungal biomass present in the particular soil. The other media used are Martin's rose bengal agar, Czapek-Dox's agar (modified) and Saboraud's agar media.

- Surveys indicate that certain suites of fungi are characteristic of particular vegetation types or geographical areas. For example, species of *Mortierella, Penicillium, Mucor* and *Qidiodendron* are more common in temperate to high latitudes, especially in forest soils.
- Species of *Aspergillus* are more common in grasslands and deserts, and at low latitudes.
- *Fusarium, Papulaspora* and *Periconia* are characteristic of grassland soils.

The dominance of most soils by fungi is impressive, whether measured by number of species, reproductive propagules (often 10^4 to 10^6 colony forming units per gram dry weight of soil), hyphal lengths (commonly 100-1000 meters of hyphae/g dry weight of soil) or biomass (37-184 g dry weight of mycelium per sq. metre).

1. The conventional dilution plate count technique reveals the presence of 104 to 106 fungal propagules/gram of soil. The propagule being

considered as any spore, hyphae or hyphal fragment that is capable of giving rise to a colony.

Furthermore, the readily sporulating genera appear in large numbers on the agar plates because each individual spore can give rise to a colony. Therefore, the profusely sporulating fungi such as *Penicillium* and *Aspergillus* are isolated frequently.

2. Microscopic visualization: Other methods used for bacterial enumeration are also used in the study of fungal population. Recent use of improved fluorescent strains like calcofluor white and fluorescein diacetate (FDA), fluorochrome [5-(and 6-) sulfofluorescein diacetate], confocal laser microscopy and digital image analysis shows promise in direct microscopic visualization of hyphae in soil smears.
3. Rossi-Cholodny buried slide technique is often used to see the mycelial growth and microbial interactions in the soil.
4. Selective inhibition of bacteria or fungi with antibacterial (streptomycin) or antifungal (cycloheximide) antibiotics had been used to determine each group's contribution to the total microbial biomass. Using this technique, estimates of the fungal proportion of the microbial biomass (measured by response to glucose added to the antibiotic-amended soil) range from 11 to 90%. The fungal hyphae accounts for about 500 to 5000 kg biomass/ha of surface soil, considering average fungal hyphae having 5 μm diameter, specific gravity of 1.2 and taking the average length of 10 to 100 metre.
5. Among the indirect method of assessing fungal biomass and activity in soils, signature biomarkers, including both nucleic acids and certain fatty acids and phospholipid fatty acids, offer various levels of taxonomic discrimination within the fungi.

Factors affecting growth of fungi in soil

All the environmental factors that affect growth and distribution of bacteria and actinomycetes also affect fungal flora in soil.

Organic matter content

The fungi are heterotrophs and therefore quality and quantity of organic matter has a direct effect on fungal population. When fresh organic crop residues are added to the soil, the fungi compete much faster because of their ability to degrade carbonaceous compounds - cellulose and hemicellulose, which are poor in nitrogen content. Their numbers and activities start declining as the C:N ratio of the organic residue becomes narrow.

The important species, which are affected greatly by addition of organic residues, are *Penicillium, Aspergillus, Trichoderma, Fusarium* and *Mucor*. The response of individual species varies with the composition of added organic residues.

Soil pH

Fungi are dominant in acidic soils whereas acidic environment is unfavourable for the growth of bacteria or actinomycetes. The fungi could grow at pH as low as 2.0 to 3.0 and could also exist at pH as high as 9.0. Sensitivity to pH could be of profound importance to soil-borne plant pathogens. Pathogens such as *Plasmodiophora brassicae* (which causes downy mildew of crucifers) fare best in acidic habitats and the disease produced by it is uncommon or mild in soil of pH greater than 7.5. Some plant pathogenic fungi grow optimally in soils near neutralitywhereas certain species are prolific in alkaline habitats. In these fungi, acidification of soil may become a practical control measure.

Addition of inorganic fertilizers lowers down the soil pH due to microbial oxidation of ammoniacal fertilizers to nitric acid and thus indirectly affect fungal population. Liming of acidic soil increases soil pH, favouring more growth of bacteria and actinomycetes.

Moisture level

Moisture level in soil affects soil aeration status. Most of the fungi are aerobic. Therefore, excess water logging and moisture decrease the fungal populations because diffusion of O_2 necessary for aerobic metabolism of fungi is inadequate to meet the microbiological demand and the fungi are among the first to suffer. Improvement in the moisture status of the environment favours fungal numbers so that, at suboptimal water levels, the count is often positively correlated with moisture.

Table . Changes in population of fungi in grassland at various moisture levels.

Average % soil moisture	Fungi x 10^3			Spores as % of total
	Spores	Hyphal units	Total	
8.9	39	60	99	49
11.2	32	57	89	36
18.5	29	113	143	20
24.2	16	133	149	10
27.1	20	153	173	12

Arable soils contain abundant fungal population because of ready oxygen diffusion. Under adverse conditions in excess wet or dry environments, fungi sporulate and persist in soil till favourable conditions return.

Temperature

Fungi are mostly mesophilic in their temperature relationships although thermophilic species are also found abundantly in decomposing composts. The thermophilic species disappears from composts that attain higher temperatures (thermophiles normally multiply at 50^0C and sometimes at 55^0C but not at 65^0C). Seasonal variations also influence the population of different temperature tolerant species of fungi.

- Among the terrestrial thermophiles are species of *Aspergillus, Chaetomium, Humicola* and *Mucor*.
- Certain species of Cylindrocarpon, Mucor, Penicillium and Cladosporium could grow as psychrophiles and grow on agar plates incubated at 6^0C.

Objective type questions

A. Fill up the blanks with suitable words:

i. ____________________ is an example of organic matter decomposing fungi.

ii. The branch of science that deals with the study of fungi is called ____________.

iii. Mycology comes from the Greek words ___________ & ____________.

iv. According to body mass of microorganisms, _____________ constitutes highest body mass in soil.

v. ____________________ is common edible mushroom.

vi. The production of antibiotics by fungal species in ascomycete genera such as _____________ & _____________.

vii. *Agaricus bisporus* are prized for their excellent flavour and large protein content varying from____ to______.

viii. *Puccinia graminis tritici* causes the disease in wheat named as ___________________.

ix. *Ustilago tritici* causes the disease in wheat named as _________________.

x. The association of specific algae or cyanobacteria with fungi results in symbiotic association called as ______________.

B. Multiple choice questions

i. The vegetative body of a fungus is called as-

a. Thallus
b. Hyphae
c. Mycelia
d. Mycelium

ii. Which of the following organism constitute highest body mass in soil

a. Bacteria
b. Fungi
c. Actinomycetes
d. Viruses

iii. Which of the following organism is causative agent of take-all disease of wheat

a. *Gaeumannomyces*
b. *Rhizoctonia*
c. *Verticillium*
d. *Sclerotium*

iv. Which of the following is an example of ectomycorrhizae -

a. *Amanita*
b. *Glomus*
c. *Acaulospora*
d. *Gigaspora*

v. Which of the following fungi are used as biopesticides

a) *Beauveria bassiana*
b) *Verticillium lecani*
c) *Trichoderma*
d) All of these.

vi. Which of the following media are used for isolation of fungi from soil

a) Martin's rose bengal agar
b) Czapek-Dox's agar
c) Saboraud's agar media
d) All of these

vii. The population of fungi in healthy soil varies between

a. 10^4 to 10^6 cfu/g soil
b. 10^5 to 10^7 cfu/g soil
c. 10^7 to 10^8 cfu/g soil
d. 10^8 to 10^9 cfu/g soil

viii. The suitable soil pH for growth of fungi varies between

a. 4.5-6.0
b. 6.0-7.0
c. 7.0-7.5
d. 8.0-8.5

ix. Which of the following is example of thermophilic fungi

a. *Aspergillus*
b. *Chaetomium*
c. *Humicola*
d. All of these

x. Which of the following is example of psychrophilic fungi

a. Cylindrocarpon   b. Penicillium

c. Mucor   d. All of these

Answer keys

A. Fill in the blanks

i. Aspergillus, **ii.** Mycology, **iii.** Mycos and Logos, **iv.** Fungi, **v.** *Agaricus campestris*, **vi.** *Penicillium* and *Fusarium*, **vii.** 21-30%, **viii.** Black stem rust of wheat, **ix.** Loose smut of wheat, **x.** Lichens.

B. Multiple choice questions

i. **a** ii. **b** iii. **a** iv. **a** v. **d** vi. **d** vii. **a** viii. **a** ix. **d** x. **d**

C. Descriptive type questions

i. Write in brief the factors affecting fungal population in soil.

ii. Classify fungi according to taxonomic classification with suitable examples for each calss.

iii. Write in brief the functions and activity of fungi.

iv. Define mycorrhizae. Classify mycorrhizae with suitable examples for each group. What are the function of vesicular arbuscular mycorrhizae in soil?

5

Algae

Introduction

Soil algae are not as numerous as bacteria, actinomycetes and fungi. Furthermore, these are photosynthetic microorganisms. Recent work has led to a more complete knowledge of the ecology and importance of terrestrial algae. These algae are found abundant in the habitats in which moisture is adequate and light is accessible.

Classification

Soil algae are classified into four major classes:

1. Cyanophyceae (Blue green algae)
2. Chlorophyceae (Green algae)
3. Bacillarophyceae (DIatoms)
4. Xanthophyceae (Yellow green algae)

In addition to these four classes, Rhodophyceae (Red algae) and Phaephyceae (Brown algae) occur less frequently and found in marine habitat. Euglenophyceae found in fresh water.

Blue green algae (Cyanobacteria) and its characteristics

1. Blue green algae are prokaryotic in cellular organization and therefore, classified with bacteria.
2. The blue green bacteria possess chlorophyll ‘a’ in membrane bound thylakoids that are distributed throughout the cytoplasm and carry out photosynthesis.
3. They also possess specific pigments such as phycobillins, a blue pigment phycocyanin and phycoerythrin (red), which are absent in eukaryotic algae.
4. A unique property of blue green algae is their ability to fix nitrogen from the atmosphere, which has so far not been detected in eukaryotic algae. Thus, blue green bacteria are photoautotrophic diazotrophs.

5. Other photoautotrophic diazotrophic bacteria like *Rhodopseudomonas* and *Chromatium* differ from the blue green bacteria as the former are anaerobic and do not evolve O_2 during photosynthesis.

6. The storage product is cyanophycean starch (the amylopectin part of starch) and proteins.

7. The blue green bacteria are unicellular as well as filamentous. Both the forms often produce mucilaginous sheath around the cells like capsular material found in bacteria. Mucilaginous material surrounding the cells can also serve as barrier to O_2 diffusion. For example, *Nostoc cordubensis* are characterized by mucilaginous colonies.

8. Filamentous blue green bacteria produce terminal or intercalary heterocysts which are specialized thick walled cells to block diffusion of O_2 into the cytoplasm for protection of nitrogenase. Both heterocysts and mucilage must be present for N_2 fixation to occurs at O_2 levels above 20%. The O_2-evolving portion of photosynthetic apparatus (photosystem II) is absent from the heterocysts, thus eliminating the biochemical production of O_2 during photosynsthesis.

9. The nucleus lacks membrane similar to bacteria.

Taxonomically, the blue green bacteria are classified into class Cyanophyceae. The major orders of class cyanophyceae are:

1. Chroococcales: Unicellular or colonical, multiplication by binary fission or by endospores, non-heterocystous. e.g. *Chroococcus, Microcystis, Placoma, Entophysates.*

2. Chamaesiphonales: Unicellular or colonical, multiplication by endospores or exospores, non-heterocystous. e.g. *Dermocarpa, Stichosiphon, Chamaesiphon, Endonema, Sophononema.*

3. Pleurocapsales: Filamentous, non-heterocystous, multiplication by endospores. e.g. *Pleurocapsa, Hyalla, Solentia.*

4. Nostocales: Filamentous, with or without heterocysts, multiplication by hormogonia (short motile filaments). e.g. Oscillatoria, Anabaena, Anabaenopsis, Phormidium, Calothrix, Tolypothrix, Dichothrix, Brachytrichia, Gloeotrichia, Scytonema, Nostoc.

5. Stigonematales: Filamentous, heterocystous, multiplication by hormogonia or by akinetes. e.g. *Pulvinularia, Capsosira, Masticocoleus, Nostochopsis, Fischerella, Haprosiphon, Weistiella.*

Environmental influences

- The blue green bacteria are essentially photoautotrophs and a majority of them have the ability to fix atmospheric N_2. These microorganisms are therefore nutritionally least dependent on soil fertility or soil organic matter.
- Some species could adapt to heterotrophic nutrition, under dark conditions. e.g. species of *Tolypothrix, Anabaena, Chlorogloea* and *Anabaenopsis*. But the growth is not as vigorous as under light conditions.
- The adequate light and moisture are the major requirement for these microorganisms. Therefore, their abundance is restricted to upper soil layers exposed to sunlight. In lower horizons of soil, only unicellular forms or spores may be isolated.
- They have a wide range of soil adaptability and are found in arid deserts, barren rocks and fresh water lakes.
- Most of blue green bacteria are mesophilic. Some could grow as psychrophiles. e.g. *Anabaena flosaque* and *Nostoc commune*. Some blue green algae are thermophilic. e.g. *Synechococcus*.
- They are abundant in neutral to alkaline soils. As a consequence, the blue green bacteria are often recommended for reclamation of saline 'Usar' lands under flooded conditions.

Eukaryotic algae

1. Eukaryotic algae are photoautotrophs and thus require access to sunlight and adequate moisture for their growth.
2. Their proliferation is normally restricted on the soil surface, although isolation could also be made from subsoil mainly because of penetration of filaments in the subsoil or passage of unicellular algae through seepage of flooding water and cultivation practices.
3. The algae are both unicellular as well as filamentous.

Taxonomically, soil eukaryotic algae are divided into:

- Chlorophyta (Green algae)
- Bacillarophyta (DIatoms)
- Xanthophyta (Yellow green algae)

Environmental factors effects

- Being autotrophic in nutrition, their growth is independent of soil organic matter. However, they require inorganic salts of nitrogen, phosphorus, potassium, magnesium, sulphur, iron and other micronutrients for growth.
- Some members of chlorophyta can adapt to heterotrophic nutrition and can use starch, sucrose, glycerol, acetone and tricarboxylic acid cycle intermediates in absence of light. But under such conditions, their growth is poor and can adapt to phototrophic nutrition as soon as they are exposed to light.
- They require adequate moisture and sunlight for growth.
- Moderate estimates show algal population in soil range from 100 to 10,000 cells/gram of soil.
- Biomass estimates show 10-50 kg of algae filaments per hectare of soil.

Chlorophyceae

1. The members of chlorophyceae are about 700 species and they possess green chlorophyll a, b in their chloroplasts. they also contain xanthophylls and carotene pigments.
2. They are unicellular as well as filamentous. Unicellular motile green algae is *Chlamydomonas*, unicellular non-motile green algae are: *Volvox, Chlorella, Spirulina, Porphyridium*.
3. In general, the cell walls of eukaryotic algae are similar to those of higher plants (cellulose). Their DNA is localized within minutely perforated nuclear membranes. In addition to the nucleus, eukaryotic algae cells may also contain cell organelles including vacuoles, flagella, golgi bodies and mitochondria. The carbon reserve material is starch and sucrose.
4. They are prominent in slightly acidic environments but could also be found in neutral and alkaline soil.
5. Widely encountered species in soil are *Chlamydomonas, Chlorella, Characium, Chlorococcus, DActylococcus, Hormidium, Protococcus, protosiphon, Scenedesmus, Spongiochloris, Stichococcus* and *Ulothrix*.

Euglena is although a member of chlorophyta but often can adapt to heterotrophic nutrition under dark conditions. The possession of photosynthetic metabolism relates them to algae but these microorganisms also resemble the non-chlorophyll containing protozoa. *Euglena* is also classifies as one of the genus of Euglenophyta.

Bacillarophyceae

- The diatoms are unicellular or colonical (filamentous) algae surrounded by a highly specified outer layer.
- Terrestrial diatoms are generally smaller than aquatic species, and most are capable of movement.
- The cell wall or frustule is composed of two slightly overlapping halves. The cytoplasm contains a single nucleus and one to many plastids.
- The chloroplasts contain chlorophyll a and c, giving the cells their characteristic color that can vary from green or yellow to golden brown. The major carotenoid is golden-brown fucoxanthin.
- The normal method of reproduction is asexual by division of one cell into two.
- The storage product in the diatoms is chrysolaminarin, a β-1,3 linked glucan.
- Diatoms form resting spores with thick, ornamented walls.
- The small size of diatoms may be advantageous since it permits greater water and salt absorption because of greater surface:volume ratios of cells.
- Diatoms are generally less frequent in acid soils and grow best near neutrality or at slightly alkaline conditions.
- The more predominant soil genera are: *Acanthes, Amphora, Caloneis, Cymbella, Fragileria, Hantzschia, Navicula, Pinnularia, Stauroneis* and *Synedra*.

Xanthophyceae (Yellow-green algae)

- Yellow-green algae are relatively rare in soil. Their chloroplasts contain chlorophyll a but lack fucoxanthin.
- Other carotenoid pigments provide the characteristic yellow-green coloration of this group.
- The cell walls of the yellow-green algae are usually composed of cellulose.

- Their principal storage product is probably a β-1,3 linked glucan but lipids are also important for food storage.
- Members of this class usually multiply asexually by fragmentation and produce motile or non-motile spores.
- Sexual reproduction has been found in *Botrydium* and *Vaucheria.*
- They have the ability to form resting spores.
- The predominant soil genera are: *Botrydiopsis, Botrydium, Bumilleria, Bumilleriopsis, Geobotrys, Heterothrix, Heterococcus, Pleuochloris* and *Vaucheria.*

Dinoflagellates: Unicellular and flagellated organisms. Possess chlorophylls a and c, xanthophylls. Storage material is starch (α-1,4-glucan). Cell wall is made of cellulose. These are mainly marine species. The predominant genera are *Gonyaulax* and *Pfiesteria.*

Function and activity of algae in soil and other ecosystems

1. Algae are photoautotrophic and generate organic matter from inorganic substances in the presence of adequate moisture and sunlight, thus increasing organic carbon of the cultivated land.
2. They play an important role in colonizing barren or eroded land surfaces and contribute to soil development in ecosystems such as deserts because they produce ecologically significant amounts of organic matter.
3. The algae also play an important role in corroding and weathering of rocks as they can grow on barren rocks on which other plants fail to grow.
    i. The weathering of rocks may be the result of carbonic acid formation from the respiratory CO_2 release of the algae and its subsequent reaction with water or
    ii. It may be associated with the metabolic products i.e. organic acids, of the bacterial and fungal utilization of the organic matter supplied by the algal protoplasm.
4. Filamentous algae helps in maintaining soil structures. Under humid environmental conditions, some species form s surface bloom on the soil, which reduce soil erosion losses as the algal bloom or polysaccharides binds soil particles tightly to form aggregates.

5. The algae evolve O_2 during photosynthesis and contribute O_2 to the submerged root environment of crop plants for respiration, especially in flooded rice fields.
6. Food source
   - *Chlorella*, a unicellular green algae, possesses a high quality of food value. It has about 50% protein and 20% of lipid and carbohydrates. *Chlorella* protein contains all amino acids essentioal for human nutrition. Besides, it contains vitamins A, B, C, K and various other essential nutrients.
   - *Ulva* is also processed as food product. *Ulva lactuca* has been formerly used in salad and soup in Scotland. Vitamins B and C are found in *Ulva*.
   - *Porphyra tenera* is most important edible red algae cultivated commercially in Japan. It is rich in proteins (30-35%), carbohydrates (40-45%) and good source of vitamins A, B, C and niacin.
7. Antibiotics:

   The genus Chlorella yields an antibiotic, chlorellin (obtained in 1944) which is used against Gram +ve and Gram -ve bacteria especially *E. coli, Shigella dysenteriae* and *Staphylococcus aureus. Ascophyllum, Laminaria, Lyngbya, Cladophora, Polysiphonia* yield antibiotic like substances.
8. Pathogenic: Cephaleuros virescens (Green algae) causes red rust on tea leaves. C. coffeae causes red rusts on the leaves of coffea.
9. Toxins: Produced by *Prymnesium parvum, Microcystis, Gonyaulax* can lead to death of fishes and aquatic animal. *Anabaena* and *Microcystis* cause gastric troubles when consumed by humans with drinking water. *Gymnodinium brevis* produce respiratory disorders whereas Lyngbya and Chlorella can cause skin infections.
10. Blue green algae substantially contribute fixed N_2 to the growing plants. Estimations show that about 20-50 kg per hectare of fixed N_2 is contributed by BGA in rice fields per growing season of the crop. Thus BGA have the potential to be used as biofertilizers; however, their application is not widespread.
11. Algae also possess the capacity to utilize the inorganic nitrogenous compounds readily and therefore, could prevent N_2 losses due to leaching and volatilization when such fertilizers are applied in standing crops.

12. Blue green bacteria form symbiotic associations with a wide variety of fungi, bryophytes, ferns, gymnosperms, angiosperms and protozoa.

    a. Fungal-algal associations form lichens. For example, species of *Calothrix, Nostoc, Gloecapsa, Scytonema.*

       Examples of N_2-fixing lichens include: *Collema, Nephroma, Peltigera, Solorina, Stereocaulon, Sticta, Placynthium, Lobaria, Lichina, Lepogium, Epheba* and *Dendriscolaulon.*

    b. *Nostoc* and *Anabaena sp.* with bryophytes - *Anthoceros, Blasia, Cavicularia.*

    c. *Anabaena* associates with fern *Azolla.*

    d. *Nostoc* and *Anabaena sp.* associate with Gymnosperms - *Bowenia, Cycan, Divon, Stengeria* and *Zamia.*

    e. *Nostoc sp.* - associate with *Angiosperms, Gunera.*

    f. *Cyanocyta, Cyanophera* and *Skujapelta* species associate with protozoa - *Cryptella, Paulinella* and *Pelina.*

13. Cyanobacteria are abundant in neutral to alkaline soils and have been used to reclaim saline soil in India.

14. Genetic engineering may allow the production of cyanobacterial strains that have an enhanced capacity for N_2 fixation or herbicide resistance. Thus offering a superior organism for bio-fertilization.

Other applications

15. Waste water treatment has become both an ecological as well as economic necessity and the commonly used method is activated sludge process (ASP). The procedure utilized O_2 produced by the phototrophic microalgae for complete mineralization or oxidation of organic matter. The method is simpler, easy and economical; and leads to the production of biomass which can be utilized in pisciculture or as an additive to poultry feed or used as fertilizer.

16. Successful operation of algal remediation system depends on establishing a dynamic equilibrium between photosynthetic O_2 production by algae and cyanobacteria; and the respiration based oxidative degradation mainly brought by bacteria like *Pseudomonas, Flavobacterium* and *Achromobacter*. Estimates made at optimum activity stage put the algae : bacteria ratio between 3:1 to 3:2 on biomass basis and 1:1000 on numerical basis.

17. **Algae as indicators of pollution:** Some algae have the unique distinction of abounding polluted habitats. For example, Microcystis, Oscillatoria, and Navicula are regarded as bioindicators of pollution.

    Others are *Chlorella, Scenedesmus, Chlamydomonas, Micractinium* and Diatoms are also capable of growing in highly polluted water. This property has given them the recognition of bioaccumulators, scavenging heavy metals and the toxic substances from the ecosystem.

    Indicator algae in the polluted habitats need to be preserved, as they are likely to be eliminated because of their constant exposure to pollutants. Regression analysis also helped in identifying algal forms, which could be treated as the role models of ecological indicators.

These forms are:

Spirulina laxissima - which existed in slightly alkaline and low conductivity experiments.

Aphanothece nidulans - which preferred slightly acidic and ammonium-rich habitat.

Oscillatoria angustissima - which dominated high organic carbon and eutrophic conditions.

Synechococcus elongates - which grew happily in presence of low available phosphorus and sulfate content.

18. **Pesticide removal:** A variety of pesticides like organochlorine, carbamates, organophosphorus and synthetic pyrethroids are now in use. Algae have tremendous capacity to absorb and accumulate organochlorines, because of their larger surface (to adsorb/absorb), which has led to their use in polluted water for treatment of industrial effluents.

    Cyanobacteria, in general, not only resist very high concentrations of organo-phosphorus pesticides, but also metabolize and use them in phosphorus nutrition. Since the residual effects of these agrochemicals are detrimental to the ecosystem, it is very essential to develop pesticide resistant/degrading strains.

Table 1. Accumulation of insecticides by algae.

Insecticide	Algae	Insecticide conc. in the medium	Bioconcentration
Aldrin	*Anabaena cylindrica* *Anacystis nidulans*	1 ppb 1 ppb	129 990
DDT	*Anabaena sp.* *Scenedesmus sp.*	1 ppm 1 ppm	268 134
Methoxychlor	*Chlorella pyrenoidosa* *Scenedesmus sp.*	0.008-0.05 ppm 0.01-1 ppm	8400 37
Hexachlorohexane (γ-HCH)	*Chlorella pyrenoidosa* *Chlamydomonas sp.*	0.01-1 ppm 0.01-1 ppm	153-267 310
Parathion	*Anacystis nidulans* *Scenedesmus obliquus* *Euglena gracilis*	1 ppm 1 ppm 1 ppm	50 72 62

19. **Heavy metal removal:** Selective uptake and accumulation of heavy metals, toxic carcinogens and other radioactive compounds by algae has gained significance in the present context of legislation on pollution abatement and effluent treatment. Removal of metals by algae can also be used to recover valuable elements such as gold and silver after appropriate treatment of the loaded biomass. This process is highly efficient especially for dilute external concentrations. Some economical processes of microorganisms-mediated metal removal are being used commercially in mining and metallurgical operations.

    Inside the cells, metal ions may be compartmentalized as in the case of *Plectonema boryanum* for cadmium and *Anabaena cylindrica* for aluminium or they can accumulate ions in polyphosphate bodies. Alternatively, they can be converted to more innocuous forms by binding or precipitation. For example, *Synechococcus sp.* synthesize extracellular polymer to bind nickel and the cell interior becomes highly granular.

    Some eukaryotic algae produce metal binding prioteins such as *Scenedesmus acutiformis* (for copper) and *Chlorella pyrenoidosa* (for cadmium). The algal biomass with adsorbed metals may be harvested for extraction. Some economical processes of microorganism-mediated biomining have already been exploited commercially.

20. **Production of unconventional energy:** Production of hydrogen energy using specific microalgae or some photosynthetic bacteria has now crossed pilot plant stage. *Chlamydomonas, Chlorella, Mastigoclades laminosus* and *Plectonema boryanum* are known to be efficient producers of hydrogen under favourable conditions. However, a lot of work need to be carried out to make the systems operational.

21. **Algae in space:** Certain species of soil algae are potentially useful for food or energy during space travel. Although the production of photoautotrophic algae on a large scale is uncommon, some have suggested that algae could be a major source of future dietary protein, both in space and on earth. Indeed, some (e.g. Spirulina) are already sold as dietary supplements.

22. **Other effects:** *Oscillatoria, Spirogyra* and diatoms clog the water filters and change the taste and odour (foul) of water. *Anacystis* and *Chaetophora* corrode the concrete and metallic walls of pipes and biolers by secreting carbonic acid, oxalic acid, silicic acid. Algae contamination causes change in pH, O_2 content and bicarbonate content of water. Agar-agar (solidifying agent used in medium preparation) is extracted from red algae, *Gelidium*.

Objective type questions

A. Fill up the blanks with suitable words

i. ________________________ is an example of symbiotic nitrogen fixing algae.

ii. *Euglena*, an actinomycetes is classifies as one of the genus of ________________________.

iii. The algae Diatoms falls under the class of ________________________.

iv. The example of marine species algae is ________________________.

v. *Plectonemaboryanum* couldaccumulatetheheavymetal_______________ in their polyphosphate bodies.

vi. Agar-agar (solidifying agent used in medium preparation) is extracted from red algae,

vii. *Anabaena cylindrica* can accumulate ___________ ions in polyphosphate bodies.

viii. *Anabaena* associates with fern ______________.

ix. *Nostoc* and *Anabaena sp.* associate with Gymnosperms _________ &_________.

x. Fungal-algal associations are called as __________.

B. Multiple choice questions

i. Majority of the eukaryotic algae are -

a. Photoautotrophs
b. Photoheterotrophs
c. Chemoautotrophs
d. None

ii. Yellow green algae are also known as -

a. Xanthophyceae
b. Chlorophyceae
c. Cyanophyceae
d. None.

iii. Blue green algae are _____________ in cellular organization

a. Prokaryotic
b. Eukaryotic
c. Mesokaryotic
d. Either b or c.

iv. Phycobillins, a blue pigment is present in which of the class of algae

a. Cyanophyceae
b. Rhodophyceae
c. Chlorophyceae
d. None

v. Phycocyanin, a red pigment is present in which of the class of algae

a) Rhodophyceae
b) Cyanophyceae
c) Chlorophyceae
d) None

vi. Cyanobacteria are abundant in neutral to alkaline soils and are used for reclamation of

a) Acid soil
b) Acid sulphate soil
c) Saline soil
d) Alkaline soil

vii. The *Chlorella*, a unicellular green algae, possesses a high quality protein amounting

a. 20%
b. 30%
c. 40%
d. 50%

viii. *Porphyra tenera* is most important edible red algae cultivated commercially are rich source of

a. Vitamin a
b. Vitamin c
c. Both a and b
d. None

ix. Biomass estimates show algal filaments weighs between ------------ to ------------per hectare of soil

a. 0-5 kg
b. 5-10 kg
c. 10-50 kg
d. None

Answer keys

A. Fill in the blanks

i. Anabaena, **ii.** Euglenaceae, **iii.** Acillarophyceae, **iv.** *Gonyaulax,* **v.** Cadmium, **vi.** *Gelidium,* **vii.** *Aluminium,* **viii.** *Azolla,* **ix.***Cycan, Zamia etc.,* **x.** Lichens.

B. Multiple choice questions

i. **a** ii. **a** iii. **a** iv. **a** v. **a** vi. **c** vii. **d** viii. **c** ix. **c**

C. Descriptive type questions

i. Write in brief the characteristics of blue green algae.
ii. Classify eukaryotic algae with suitable examples for each groups. Briefly describe the characteristics of bacillariophyceae.
iii. What are the function and activity of algae in soil and other ecosystems?
iv. What are the environmental influences on the population and activity of algae in soil?
v. Write in brief the application and uses of algae in human welfare.

6

Protozoa

Introduction

The soil protozoa are unicellular and the simplest forms of animal life. Most of the protozoa are devoid of chlorophyll but transitional genera resemble the algae and possess chloroplasts containing chlorophyll pigments. The active life stage is commonly known as trophozoite phase where they feed and multiply but under adverse climatic conditions, they form resting stage or cyst, which develop a thick coating on the cell surface. Mostly, the protozoa multiply by longitudinal or transverse fission but some species reproduce sexually by fusion of two cells followed by exchange of genetic material between their nuclei and cell division.

Taxonomic classification of soil protozoa is based on their means of loacomotion. The Mastigophora or flagellates move about by means of one to four or more long flagella. Their cell size varies from 2 to 50 µm in length. Some members of flagellates possess chlorophyll. The important genera of terrestrial habitats are *Allantion, Bodo, Cercobodo, Cercomonas, Entosiphon, Heteromita, Monas, Oikomonas, Sainouran, Spiromonas, Spongomonas* and *Tetramitus*. The chlorophyllous form, which are often referred as algae as well as protozoa, are members of *Euglena*. Only non-chlorophyll bearing flagellates occur in soil; the photosynthetic species are strictly aquatic.

The second category of soil protozoa is members of class Sarcodina (rhizopdes or amoebae), which move by means of pseudopodia - temporary protoplasmic extrusions from the cell body. Their cell size varies from 2-600 µm in length. The representative genera are *Amoeba, Acanthamoeba, Biomyxa, Diffugia, Euglypha, Hartmanella, Lecythium, Naegleria, Nuclearia* and *Trinema*.

The soil amoeba or sarcodina are of two types: testate amoeba have a shell (test) like structure, others possess none. When the shell is present, the pseudopodia extend through distinct openings. The testate amoeba live within a test (or shell) constructed of soil particles bound together by secretions. The size and shape of the test, the size and shape of the scales and spines or horns on the surface of test, and the placement and border pattern of the pseudostome or "false mouth" (the opening in the test that allows the amoeba to go in and out of its "mobile home") are important characteristics for identifying species.

The third category of terrestrial protozoa moves by means of cilia and classified as Ciliophora or ciliates. Their cell size varies from 50 to 1500 µm in length. The whole body is covered with small cilia. The important genera are Balantiophorus, Colpidium, Colpoda, Enchelys, Gastrostyla, Halteria, Oxytricha, Pleurotricha, Uroleptus and Vorticella. Distribution of cilia, fusion of cilia into tufts called Cirri or into membranelles (undulating membranes) and the placement of cytostome (mouth) with associated membranelles are important features for species identification.

Nutrition

The photosynthetic algae-like protozoa synthesize their protoplasm from CO2 using energy derived from sunlight (photoautotrophy). Nutrition of the terrestrial protozoa is either saprobic obtaining nutrition from soluble organic and inorganic subsytances of phagotrophic - direct feeding on live or dead microbial cells and organic particles. The phagotrophic nutrition is more common among soil protozoa. Mostly, the protozoa feed on bacteria and other microorganisms but some are cannibalistic feeding on cells of their own species. There is no absolute specificity with regard to phagotrophic nutrition of protozoa but certain species may prefer bacteria and some other preferably feed on algae. For example, isolates of *Acanthamoeba, Hartmanella* and *Mayorella* feed preferably on bacteria and yeast, while isolates of *Amoeba* and *Tetramitus* feed on algae.

Distribution

In soil, protozoan population range from 10^4 to 10^5 cells per gram of soil and is directly related to bacterial and other microflora population. The biomass of protozoa in a moderately fertile soil may range from 5 to 200 kg per hectare of soil. Protozoa occurs in all ecosystems of the world, from the poles to the tropics, from deserts to humid jungles and from the tops of mountains to the depths of the ocean.

Environmental factors affecting protozoan population

The impact of various environmental and land management factors on protozoa can be understood in terms of their effects on substrate supply and secondarily on soil structure, moisture and climatic factors.

Organic matter

Protozoa tend to be more numerous at sites of decomposing organic matter than in the rhizosphere because of increased concentrations and absolute amounts of substrate. A 90-fold increase on decomposing grass residues in soil has been demonstrated. The application of organic manures to soil is an obvious source of added substrate and consequently increases protozoan populations. The application of inorganic fertilizers generally increased protozoan populations, because of the increased plant growth, which led to a greater organic matter input in the soil. A transition from grassland (permanent plant cover) to arable land (annual cultivation) reduced the energy input to the soil and consequently protozoan biomass.

Soil depth

The characteristic decrease in the number of protozoa with soil depth, from approx. 10^6/gram at 10 cm to 10^2/gram below 1 meter is likely to result from decreasing levels of organic matter down the soil profile. Protozoa can survive at considerable depths and viable cells have been recovered from 7.4 metres below the surface.

Moisture

Adequate moisture is essential for the vertical and lateral movement. Under moisture stress conditions, protozoa encyst and remain in cyst form until environment becomes favourable for growth. As protozoa require a water film for locomotion and feeding, their activity is limited to the water-filled pore space in soil.

pH

With regard to acidity, most protozoa exhibit no marked sensitivity to pH. Certain species can proliferate at pH 3.5 and others at above pH 9.0. On the other hand, many protozoa do not tolerate extreme acidity or alkalinity and fare poorly outside the range from pH 6.0 to 8.0. The application of elemental sulphur to a Canadian pasture, however, increased acidification of the soil from pH 5.7 to 44.7 and so reduced number of soil microorganisms and mycophagous protozoa.

Temperature

Temperature is another ecological determinant. The most favourable environments are being cool and damp environment. Excessive warmth is detrimental.

Significance in soil

1. Regulation of bacterial population in soil

   The protozoa feed on bacteria and each protozoan feed thousands of bacteria prior to cell divisions, thus protozoa helps in regulation of bacterial population. Although, the numbers and species composition of bacteria can be controlled by protozoan predation, but bacterial survival can be improved by adding clay. The added clay allows the formation of microsites in which protozoa can not reach their bacterial pray.

2. The second important function is to maintain microbial population in a balanced equilibrium. Certain bacteria, which multiply at a faster rate, could eliminate the other form that multiply at a slow rate. By uptake of fast growing bacteria, because of their ready availability to protozoa, the other forms remain in balance in nature.

3. Protozoa have a direct and indirect effect on energy and nutrient flows in the soil. Approximately 40% of the ingested nutrients are used for the production of protozoan biomass and 60% is excreted. The excreted nutrients form a direct flow of nutrients into the soil system, which are available for plant uptake and microbial growth.

   Protozoa in soil are largely responsible for an enhancement of the soil's N, P and S mineralization by their ingestion and digestion of bacteria and fungi. Moreover, they can enhance soil nitrification, immobilization of carbon, nitrogen and phosphorus in their own biomass. They further participate in soil respiration.

4. Some pathogenic protozoa such as Entamoeba histolytica which causes amebic dysentery and Naegleria which causes meningoencephalitis are soil borne and may survive for a short period as cysts in soil. Besides mammals, earthwotms, larvae of ainsects and other invertebrates may also be parasitized by soil protozoa.

5. Protozoa may inhibit the development of disease-causing organisms in soil. Soil predators in a variety of ways can suppress both bacterial and fungal pathogens, but managing these interactions for maximum effect has not yet been accomplished.

6. Protozoa serve as food for predators such as nematodes and arthropods, and large protozoa prey on small protozoa. Ciliates consume flagellates, often in preference to bacterial prey. Nematodes and large protozoa can be maintained on cultures of flagellates or amoebae in the laboratory, but the importance of these interactions in field situations is unknown.

7. Although protozoa can cause plant disease in certain circumstances, protozoa are more useful as indicators of disturbance, chemical impacts, addition of genetically altered bacteria and soil degradation.

Objective type questions

A. Fill up the blanks with suitable words

i. Soil Protozoa are ____________________ in nature.

ii. The active life stage of protozoa is commonly known as ______________

iii. pH range optimal for maximum metabolic activity of protozoa is________________.

iv. The process in which protozoa attaches themselves to a host's body is known as_____________.

v. African sleeping sickness is caused by_______________.

vi. In soil, protozoan population range from _______________cells per gram of soil

vii. Protozoa that eat other organisms are known as_______________.

viii. Red snow of high altitudes is due to the presence of_______________.

ix. When parasitic protozoa parasitize other protozoan it is known as_______________.

x. The maximum temperature that protozoa bear is__________.

B. Multiple choice questions

i. The member of subphylum Mastigophora moves with the help of -

a. Cilia b. Foot

c. Flagella d. Whip

ii. Protista differs from Monera in having-

a. Cell Wall b. Nuclear Membrane

c. Flagella d. Autotrophic Nutrition

iii. The slime moulds are characterized by the presence of-

a. Elaters b. Pseudoelaters

c. Capillitium d. All of these

iv. Cell size of Ciliophora varies from

a. 50 µm to 100 µm
b. 50 µm to 1000 µm
c. 50 µm to 1500 µm
d. 1500 µm to 2000 µm

v. Most soil protozoa are flagellates or amoebas, having their dominant mode of nitrogen as-

a) Ingestion of bacteria
b) Ingestion of fungi
c) Ingestion of mold
d) All of these

vi. Which of the following is least likely to have a rigid cell wall

a) Bacterium
b) Archaean
c) Fungus
d) Protozoa

vii. A protozoan is defined as

a. Motile prokaryotic unicellular protist
b. Motile eukaryotic unicellular protist
c. Motile eukaryotic unicellular photosynthetic protist
d. Motile eukaryotic multicellular protist

viii. Members of protozoa may motile by all methods except

a. Flagella
b. Gliding by slime secretion
c. Cilia
d. Pseudopodia

ix. Members of which phylum of protozoa are being considered as biological control agents for certain insects

a. Microspora
b. Sarcomastigophora
c. Apicomplexa
d. All of these

x. Protozoa are generally not

a. Multicellular
b. Microscopic
c. Lacking cell walls
d. Eukaryotic

Answer keys

A. Fill in the blanks

i. Unicellular **ii.** Trophozite **iii.** 6-8 **iv.** Ectocommensalism **v.** Protozoa **vi.** 10^4-10^5 vii.Holozoic **viii.** Protozoa **ix.** Hyperparasitism **x.** 40°C

B. Multiple choice questions:

i. **c** ii. **b** iii. **c** iv. **c** v. **a** vi. **d** vii. **c** viii. **b** ix. **a** x. **a**

C. Descriptive type questions

i. How the presence of protozoa in soils is significant?

ii. Write in brief about the taxonomic classification of protozoa.

iii. How soil depth and organic matter affects the population of protozoa.

iv. Please explain the mode of nutrition of protozoa?

v. How protozoa bring biological equilibrium in soils?

7

Viruses

Introduction

The microbial cells of bacteria, actinomycetes, fungi, algae and protozoa are visible by light microscopy. Beyond the resolution of light microscope are the unique organisms that are sub microscopic and intracellular parasites known as viruses. The term virus is a Latin word means poisonous substance. The viruses are functionally and metabolically different from the cellular organisms as (i) they are infectious agents and contain only one type of nucleic acid encapsulated in a protein coat. Some may possess membranes but do not have any cytoplasm or metabolism of their own. (ii) They are much smaller in size than the smallest bacteria and can pass through bacteriological filters. The diameter of the bacteriophage rarely exceeds 0.05 to 0.10 μm. The tail is somewhat larger, about 0.2 μm in length, but it is quite narrow. (iii) They are obligate parasites and multiply inside the host cell. They do not proliferate or grow in the culture media in which other microorganisms could grow under cultural conditions. (iv) The viruses are host specific and are categorized as viruses that are specific to plants, animals and microorganisms. (v) The viruses do not grow or metabolize and therefore, common antibiotics do not affect viruses. Viral diseases, however, may be treated with antiviral agents that target virus-specific components. For example, acylovir drug inhibits replication of Herpes simplex and chicken pox virus by interfering with replication of DNA. Amantadine prevents attachment of influenza virus to host cells in respiratory tract. Azidothymidine (AZT) inhibits key enzyme involved in HIV's replication cycle. Interferons produced by host cells after viral infection protect neighbouring cells against viral penetration.

Discovery of viruses: Emergence of Virology

Virus is an ultramicroscopic, disease producing, living nucleoprotein particle that can multiply only within living organisms. Intact virus unit of infection particle is called virion. Iwanowsky (1892) was the first to observe that extracts of tobacco plant with mosaic disease retained infectivity after passing through bacteria-proof filters. He concluded that infective agent is smaller than bacteria. M. W. Beijerinck (1888) observed that infective agent could diffuse into agar gel and called it as contagium vivum fluidum (contagious fluids).

Loeffler and Frosch (1898) were the first to show that there were filterable agents which could be transmitted from one infected animal to another. They reported that filtered and unfiltered lymph from animals suffering from foot and mouth disease (Rhinovirus) was similar in their infectivity because filtered lymph could subsequently serve as a source of inoculum for infecting healthy animal. It was conclude that infective filterable agent was not a toxin but an agent capable of multiplication.

Edward Twort (1915) and d'Herelle (1916) independently discovered that bacteria are susceptible to infective agent, which passed through bacteria-proof filters. These viruses are called bacteriophages. M. W. Stanley (1935) isolated the tobacco mosaic virus (TMV) and crystallized it. N. W. Piere and F. C. Bawden, British Biochemists (1936) showed that TMV contained protein and nucleic acid (single stranded RNA). Franckel-Conrat and Singer (1957) demonstrated that RNA is the genetic material in tobacco mosaic virus. In other viruses, the nucleic acid could be DNA or RNA with smallest 5 kb to largest 190 kb pairs (single stranded or double stranded, linear or circular, enveloped or naked).

Virions - Nucleocapsid virus particles that carry the genetic material capable of infecting host cells. Size varies from 0.02 to 0.3 μ depending on type of virus in size. The genetic material could be DNA or RNA; single stranded or double stranded, linear or circular, enveloped or naked, smallest 5 kilobases to largest 190 kb pairs.

Viroids - Ultramicroscopic in size, single stranded RNA molecules without any protein coat. They infect plants resulting in stunted growth and devastating effect in some plant. e.g. Coconut palms.

Prions - These are proteinaceous infectious particles and cause kuru and mad cow disease. Stanley B. Prusiner (1982) was the first to postulate their existence and won Nobel Prize in Physiology and Medicine in 1997.

Structure

The viral particles are of different size and shapes. Some are rod shaped with helical structure. Some are spherical and others appear to be combination of helical and nearly spherical structures. In rod shaped helical structure, the viral chromosome or nucleic acid is present in the head with protein subunits clustered around it. The protein subunits of the helix are termed as capsid proteins or capsomere and the entire helical structure is termed as nucleoprotein.

The spherical virions in electron micrograph are actually polyhedral as the proteins capsomeres are arranged to form a polyhedral shell, usually composed of 20 triangular faces (an icosahedron) that surrounds nucleic acid. The number of capsomeres in a virion is a characteristic property of the class to which the polyhedral virus belongs. Some virions are complex containing a polyhedral (spherical) head connected to helical structure called tail. Such structure is termed as binal structure and is relatively common among bacteriophages but not found in animal and plant viruses.

The number of proteins coded by the phage genomes varies considerably. The smallest known virus, the satellite virus of tobacco necrosis virus, codes only a single protein. Angiosperm-infecting viruses e.g. tobacco mosaic virus codes for four proteins. λ Phage codes for 46 proteins whereas bacteriophage T4 codes for 100 proteins. By comparison, the higher organism-infecting viruses, such as the pox-virus codes for a few hundred proteins.

T2 phage

T-even phages (T2, T4, T6 phages and λ Phage) have been mainly used in phage studies. The overall structure of T2 phage as revealed by electron microscope resembles that of a tadpole.

Head: It is an elongated bipyramidal, hexagonal, prism shape about 1000Å long and 650 Å wide. The head is composed of repeating identical protein monomers, which vary in size according to the phage. The nucleic acid of phages is located within the head.

Tail: It is about 1000Å long and 250 Å wide. It serves as an organ of attachment to host cells. The tail of most phages has the structure of a single protein tube, termed core (about 70Å wide), but some phages also have a thick outer tube, termed sheath, surrounding the inner core. Sheath is a hollow cylinder (about 180Å wide) built up of a number of helically arranged subunits. The sheath of T2 phage contains 24 annular rings (which form a tube) that are joined to the phage head by a thin collar.

The tube's distal end is attached to hexagonal base plate. The plate has six small spikes (termed as tail pins) and six tail fibres that are 1300Å long and 20Å wide. These tail fibres are required for adsorption to the bacterial host cell. The tail possesses the property of contractibility, which plays a role in forcing the phage nucleic acid through hollow tail into the bacterium.

Simple viruses: The virus particles are assembled in helical (rods) or icosahedral (spherical) symmetry. Those without tails and other secondary structures are referred as simple viruses.

Complex viruses: The virus particle having additional structure like head and tail are referred as complex viruses (bacteriophages). The genetic material in the complex viruses is confined in the head.

Enveloped viruses: Some animal viruses are enclosed in a phospholipid bilayer derived from host membrane in which the lipids are derived from the host cell whereas the glycoproteins impregnated in the membrane, are encoded by the virus genome.

Naked viruses: The viruses without bilayer membrane are referred as naked viruses.

Viral multiplication and reproduction in bacteriophages

Bacteriophages have two life cycles after infection on the basis of their interactions with bacterial cells i.e. lytic cycle and lysogenic cycle. Phages that cause lysis of the host cell are virulent or lytic phages. e.g. T4 and φX174 phage of *E. coli*. Other phages which enter into lysogenic relationship are temperate or avirulent phages. e.g. Lambda (λ) bacteriophage of *E. coli*.

Lytic cycle

There are five stages in lytic cycle of virulent phage.

1. Adsorption, 2. Penetration, 3. replication, 4. Assembly and maturation, 5. Lysis.

Adsorption

A typical bacteriophage gets attached on the host bacterial cell surface on a receptor site with the help of tail fibres. The base plate and spikes help to fix the phage to the cell. Virals proteins interact with bacterial cell membrane receptors i.e. proteins, LPS, polysaccharides, lipoprotein-lipopolysaccharide complexes to which the virion is attached. In some bacteria, pilli or flagella act as receptors whereas cell envelope components act as receptors in other viruses.

In animal cells, glycoproteins of RBC act as receptors for influenza virus; cell surface lipoproteins act as receptors for poliovirus. Several plant and animal viruses do not have cell receptors and these enter into the cell by phagocytosis or endocytotic processes.

Penetration

Penetration of phage after attachment to the bacterial cell differs from host to host. In bacteria, for example T4 phage penetrates through attachment of tail fibres on cell wall, which interact with core polysaccharides of outer surface. The tail fibres retract; core of the tail interacts with cell wall, release lysozyme-like enzymes and make a micro-pore. The tail sheath acts as a microsyringe and contraction of sheath results in passage of phage nucleic acid through the tip of the tail into the cytoplasm of bacterial cells. Most of the proteins of the phage (protein coat) are left outside the cell. The DNA of virus is protected by restriction enzyme action (defence mechanism of host) by glucosylation (T-even phages) and methylation of end nucleotides to prevent endonuclease activity. T3 and T7 phages encode proteins that inhibit host restriction system.

Replication

After penetration of nucleic acid, the synthesis of host cell components and cell division is stopped. The replication of bacteriophage DNA takes place. Moreover, synthesis of phage specific enzymes begins which result in the synthesis of viral-specific head and tail proteins.

Maturation and Assembly

The viral genome along with viral head and tail proteins are assembled in the host cell to form complete phage (virion) particles.

Lysis

Most of the lytic soil bacteriophages lyse their host in late stage of the infection cycle. Bacterial viruses often encode for a phage lysozyme late in the viral infection cycle that dissolves the peptidoglycan layer of the bacterial cell wall, thereby, releasing the new virions into the soil. Several hundred viruses may be released from a single cell in a lytic cycle.

Lysogeny or temperate phage cycle

λ phage - After entry into the host cell, the phage genome either enter into the lytic cycle or its replication is synchronized with host genome resulting into the establishment of lysogeny. In this cycle, the adsorption and penetration stages are similar to that of the lytic phage. After the entry of nucleic acid into the bacterial cell, the phage genome that has cohesive ends, is stably integrated into the genome of the host and continues to multiply along with the host DNA or remain independent by forming self-replicating circular DNA. The phage genome produces repressor proteins inhibiting expression of phage genes involved in lytic cycle. The host cells become lysogens and the phage is in a provirus or prophage state. The integrated phage genome is then referred as the prophage. This relationship is known as lysogeny. The bacterial cells in this relationship are known as lysogenic bacteria.

Induction of lysogenic cycle into lytic cycle

When a prophage is freed from the restrictions of lysogeny, this process is called induction. The induction of lytic cycle could be brought about by environmental changes. Occasionally, external stimuli such as starvationof the host cause transitions from lysogenic phase to lytic modes of existence. In a culture of lysogens, only a small fraction of cells undergo lysis (10^{-2} to 10^{-5} per cell). The lysogenic circuit is broken due to inactivation of the repressor molecule. The exposure of the lysogenic bacterium to UV radiation causes induction to lytic cycle. The viral nucleic acid is excised from the genome of the host and the lytic phase of the virus cycle begins. The outcome depends on the conditions of infection and the genotypes of the phage and the bacterium.

As the phage genome is integrated with the host genome, by induction sometimes the phage carries a part of the host genome, which also multiplies with the phage genome and enclosed in the virion particle during maturation. When the phage infects the other host cell, the bacterial genes attached to the phage genome is integrated with the genome of the second host bringing genetic variability by way of transduction. Lysogenic phages are therefore important tools for gene transfer of the microorganisms.

Classification of bacteriophages

Family name	Nucleic acid	Phage	Particle size
Myoviridae	Linear ds DNA	T2, T4, T6, P2	Icosahedral head, complex tail with fibres
Pedoviridae	Linear ds DNA	T3, T7	Hexagonal head, short tail
Styloviridae	Linear ds DNA	T5, λ	Hexagonal head, simple tail
Microviridae	Circular ss DNA	φX174, S13	Icosahedral head
Inoviridae	Circular ss DNA	M13, fd, fl	Filamentous
Leviviridae	Linear ss RNA	Qβ, R17, MS2	Icosahedral
Cystoviridae	Linear ds RNA	Φ6	

Plant viruses

Tobamoviruses	Linear ss RNA	Tobacco mosaic virus (TMV)
Comoviruses	Linear ss RNA	Cowpea mosaic virus (CpMV)
Potyviruses	Linear ss RNA	PYV
caulimoviruses	Circular ds DNA	Cauliflower mosaic virus (CfMV)
Geminiviridae	Linear ss DNA	Maize streak virus

Animal viruses

DNA viruses			
A. Linear ds DNA			
Enveloped		Naked	
Poxviridae	Pox viruses	Adenoviridae	Adenoviruses
Herpesviridae	Herpes viruses		
B. Linear ss DNA			
Naked			
Parvoviridae	Parvo viruses		
C. Circular ds DNA			
Enveloped		Naked	
Baculoviridae	Baculoviruses	Papovaviridae	Papovaviruses
RNA viruses:			
A. Linear ds RNA			
Reoviridae	Reo viruses (10-12 proteins, ds RNA molecules)		
B. Plus stand			
Enveloped		Naked	
Togaviridae	Toga viruses	Picornaviridae	Picorna viruses
Retroviridae	Retro viruses		
C. Minus strand ss RNA			
Enveloped			
Paramyxoviridae	Paramyxo viruses		
Rhabdoviridae	Rhabdoviruses		

Replication - Synthesis of proteins and nucleic acids.

ds DNA: One strand (-) direct transcription to mRNA (+ sense)

ss DNA: Replication to ds DNA followed by transcription to mRNA (+sense)

ssRNA + strand: direct act as mRNA, RNA replicase, - strand, copied to + strand

ds RNA: Minus strand copied to m RNA (+strand)

Isolation of bacteriophages

To determine the presence of a bacteriophage, soft-agar overlay method is used. A soil sample is incubated with the inoculation of a particular host bacterium to enrich the population of a specific phage. A small quantity of the enriched soil is added to a broth solution of the host bacterium and incubated for 24-48 hours. The broth is centrifuged and the supernatant is passed through a sterilized bacteriological filter (0.45μm size). The filtrate is then mixed with host organism and poured over hardened 1.5% agar nutrient medium. After a few days incubation, virion will lyse the host cell and create plaque formation on agar medium with lawn of host strain. Count plaque-forming units per ml of the filtrate. By this method, viruses specific to soil bacteria like *Azotobacter, Agrobacterium, Arthrobacter, Bacillus, Bdellovibrio, Clostridium, Corynebacterium, Erwinia, Mycobacterium, pseudomonas, Rhizobium, Xanthomonas* and *Cyanobacteria* have been demonstrated.

Soils usually have higher population of the phages (10^{11} viruses per gram soil) specific to different host species. On the other hand, enhancing the reproduction of host cells already in the soil to provide more individuals to parasitize will frequently lead to increases in bacteriophage abundance. Thus, addition of simple sugars, by stimulating *Arthrobacter globiformis* ultimately leads to large populations of bacteriophages capable of lysing that bacterium. Similarly, adding nutrients to soil and incubating it at higher temperatures not only favours the growth of thermopilic bacteria but also results in isolation of many bacteriophages able to lyse the thermophiles.

Detection

Electron microscopy is the most universal method to observe viruses in the soil. Immunofluorescence microscopy is more specific, viewing fluorescently labelled antibodies that fluoresce when excited with specific wavelengths of UV light while bound to the virus (i.e. antigen). The enzyme linked immunosorbent assay (ELISA) is similar to that an enzyme that acts on a specific substrate (e.g. a phosphatase) is attached to the antibody, giving a visible color reaction in the well of microtitre plate when it attaches to virus particles sorbed to the well and reacts with the substrate for the enzyme. Monoclonal antibodies allow the detection of a single trait, often chosen because it is associated with a single type of organism whereas polyclonal antibody recognizes any trait and often many different organisms. the use of gene probes also offers promise as a means for detecting viruses in soil. Use of polymerase chain reaction (PCR) to amplify low copy numbers of viral DNA or RNA may find application in the future, especially when methods of nuclleic acid extraction from the soil are improved. Theoretically, the use of PCR will allow the detection of DNA from a single viral particle.

Survival of viruses in soil

A variety of factors affect the survival of viruses in soil.

1. First is the type of virus. Some viruses are well adapted for survival in soil and persist for many years. Other viruses, such as human enteric viruses, are ill adapted for survival and may persist in soil for only days to months.
2. A second factor is the presence of a susceptible host. When populations of the host are high, populations of the homologous virus will increase over time. This shows that continued cycles of infection are occurring. Conversely, when host populations are low, then populations of the homologous virus are low since fewer encounters take place between the host and virus. As long as an appropriate host is present, a virus can persist indefinitely through continual production of new viruses.
3. A third factor affecting survival is the collective influence of soil abiotic factors such as temperature, moisture and pH. Temperature and desiccation are believed to be important factors affecting virus survival; low moisture and high temperature cause viruses in soil to rapidly lose their ability to infect a host. Soil pH affects population of streptomycete phage, with the highest numbers found in neutral to slightly alkaline soils. Certain animal and human viruses remain ineffective in soil for periods ranging upto several months. The persistence of the mouse encephalomyelitis virus, for example, is greatest in neutral environments, negligible in acid conditions and is of intermediate duration in alkaline soils.
4. A fourth factor affecting the long-term viability of phage in soil is adsorption onto organic matter and clays. This adsorption is the result of the electric double layer on clay minerals, cation exchange capacity, electrostatic interactions, hydrophobic interactions and cation bridging. Presumably, adsorption onto organic and clay surfaces protects virions from predation, degradative enzymes (e.g. extracellular proteases) and other soil biotic factors. For example, reovirus disappeared within 7.5 weeks in distikled water only, but survived for 22 weeks when kaolinite and smectite were added. Tobacco mosaic virus,, which causes a necrotic disease on tobacco leaves, is rpidly inactivated in a nonsterile soil, but is stable in an identical sterile soil.

Because of the interaction of all the factors, it is not possible to define a specific period that a given virus will persist in soil. Under harsh abiotic conditions where no host is present, viruses may survive only a few hours in soil. Alternatively, when soil is cool and moisture is present, viruses may remain infective for many months, even years. Further, if a temperate phage is incorporated into the genome of a host bacterium, its nucleic acid may persist for generations as a stable component of the ecosystem.

Importance of viruses

the overall importance of viruses in soil is difficult to determine because methods of studying viruses are imprecise. Both the number and diversity of bacterial species in soil are undetermined, so the number of bacteriophages in soil is undetermined as well. However, high number of viruses in fertile agricultural soils (10^{11} viruses per gram soil) suggests that they may play a significant role in soil ecosystems, though few viral functions have been elucidated.

1. Agriculturally important host bacteria for which bacteriophages have been reported include *Rhizobium, Bradyrhizobium, Bacillus, Arthrobacter, Pseudomonas, Azotobacter* and many species of actinomycetes. Early reports suggested that the failure of legumes to nodulate was because of the presence of high populations of a homologous bacteriophage in soil, although more recent evidence has failed to confirm these observations. Inoculation of soil with bacteriophages has been proposed as a means to reduce soil populations of undesirable and ineffective strains of Rhizobia.

2. Insect viruses have received increased attention recently because of their potential as bio-control agents. Nearly all of the economically important viruses for biocontrol are highly specific and virulent. The nuclear polyhedrosis viruses (NPV) are most often used for biological control of insects. For example, to control an agricultural infestation of celery looper, the soil could be inoculated with the *Anagrapha falcifera* multiple nuclear polyhedrosis virus (AfMNPV). The virus can potentially infect and kill all susceptible insects. When insects hosta re no longer present, the virus also disappears.

Baculoviruses are rod shaped, double stranded circular DNA (80-220 kb) viruses that can infect and kill a large number of different invertebrate organisms. In nature, some of these baculoviruses are important for the control of certain insect pests and several have been registered for use as biological insecticides. One positive feature for their use as insect biocontrol agents is that, for the most part, they have limited host ranges and do not affect non-target organisms. Nuclear polyhedrosis virus (NPV) when spread @ 250-500 l ha-1 mixed with 1% jaggery during evening hours has given promising control of polyphagous pest, *Helicoverpa armigera* in bengal gram, tomato and pigeon pea. NPV protects the cotton, maize and sorghum from infection of *Heliothis sp.* (cotton bollworm).

3. The presence of human viruses in wastewater is cause of great concern. Nearly 100 types of viruses have been detected in the wastes of humans, and these may reach the soil through land application of biosolids (sludge) and wastewater effluent from treatment plants, septic tanks and filter fields. Furthermore, the movement of these viruses through soil and the subsequent contamination of groundwater is another troubling possibility e.g. hepatitis virus.

Sewage treatment often fails to completely inactivate viruses in wastewater. Primary and secondary treatments typically remove approximately 90% of the viruses contained in the original waste. Although tertiary treatment removes additional viruses from wastewater, removal is never 100% efficient. Because theoretically only a single virion is required to initiate disease in a susceptible individual, even low numbers of viruses in wastewater represent a significant health concern. This is in contrast to bacteria, where hundreds, perhaps even millions of cells are required to initiate a bacterial infection. Attenuated (viable but not capable of causing disease) viruses may also be inoculated onto plants to protect the crop from subsequent infection by pathogenic viruses.

4. Viruses acting on fungi first attracted attention when they were identified as the causal agents of a disease of the commercial mushroom, *Agaricus bisporus*. Subsequently, it was shown that either viruses or bodies morphologically akin to viruses are present in many genera of the fungi. Among the recognized hosts are species of *Aspergillus, Boletus, Gliocladium, Fusarium, Mucor, Penicillium, Rhizopus* and *Cephalosporium* and many other genera.

The influence of the viruses on their hosts, in culture ranges from little detectable or no harm to others that bring about abnormal growth of hyphae and induce the development of abnormal fruiting bodies. Most of the viruses are entirely avirulent and do not induce lysis.

5. Viruses infecting blue-green algae are similar to those associated with the bacteria. They produce plaques on agar containing visible growth of the host algae and cause extensive lysis of algal cultures. These cyanophages resemble bacteriophages in morphology and in the process of infection. Their hosts in culture include *Anabaena, Anacystis, Nostoc, Oscillatoria, Phormidium* and *Synechococcus*.

6. The viruses also cause diseases of many agronomic and horticultural crops. But plant-infecting viruses rarely persist or overwinter in soil. These viruses are considered soil-borne only in the sense that they retain their infective capacity for some time after crop removal. Such viruses do not multiply in soil; instead, they persist for varying periods in a condition resistant to inactivation. The

viruses responsible for the mosaic disease of wheat, oats and tobacco, the big vein disease of lettuce and the corky ring spot of potatoes are soil-borne.

Plant infecting viruses rely on either nematodes or fungi for transmission. For example, tobacco rattle viruses and raspberry ring spot are spread through soil by nematodes of the genera *Longidorus, Trichodorus* and *Xiphinema* whereas tobacco necrosis and lettuce big vein viruses are disseminated by fungus *Olphidium brassicae*. The tobacco mosaic virus may sometimes endure in plants remains buried at lower soil depths for more than a year. Many of the plant viruses transmitted by nematodes are inactivated readily when soil dries whereas the plant infecting viruses borne by fungi may remain active in dry soil for many years.

7. It has been found that temperate viruses are present within most soil bacteria and that they are intimately involved in the survival and evolution of the host, sometimes adding new genes or changing existing ones. The acquisition of new traits due to lysogenization is referred to as phage conversion or lysogenic conversion. It is possible that the virus may participate in the transmission of genetic material from one bacterium to another through transduction.

This transfer could give the receptor cell physiological properties that might be of competitive advantage, such as new nutritional or biochemical attributes. Such means of genetic exchange may prove to be of importance in the evolution and variation of microorganisms.

Viral diseases

Hepatitis (liver cells), AIDS (T cells), Rabies by dog, cat and bats (Nerve cells of the brain).

Viruses as carcinogens

i. Herpesviruses associated with tumours of human cervix.
ii. human T cell leukemia virus (HTLV) causes t cell leukemia. This virus cause neurological pain disorder and destry sheaaths of nerve fibres.

Defence against viruses

Antiviral agents: Acylovir, amantadine, azidothymidine and interferons

Viral vaccines: Production of antibodies in the body's immune system.

Inactivated viruses (unable to replicate) e.g. Salk vaccine used against polio.

Attenuated vaccines (live but low replicating strain) e.g. Sabin vaccine used against polio, and to prevent chicken pox, measles, mumps and rubella.

Genetically engineered vaccine: Viral proteins are produced in yeast cells. These proteins are purified and used as vaccine. e.g. Hepatitis B vaccine.

Viral genes could be transferred and expressed in fruits such as banana.

Objective type questions

A. Fill up the blanks with suitable words

i. Viruses require____________________ for growth.

ii. When a virus enters a cell but does not replicate immediately, the situation is called ____________________

iii. The first step in infection of a host bacterial cells by a phage is________________

iv. The viral nucleocapsid is the combination of______.

v. Enveloped viruses have a __________ shape.

vi. A change from lysogeny to lysis is generally induced by________________

vii. The capsomeres consist of a number of proteins subunits or molecules called________________.

viii. The size of viruses is usually measured in________________.

ix. Viruses largely lack metabolic machinery of their own to generate energy or to synthesize______.

x. The bacterial viruses having head made up of large capsomeres, but no tail is morphologically classified as___________

B. Multiple choice questions

i. Viruses that attack bacteria are called

a. Virophage　　b. Lysophage

c. Bacteriophage　　d. None of them

ii. Viral genome inserted to the bacterial DNA is termed as

a. Lysogeny　　b. Prophage

c. Lytic cycle　　d. Virulent phage

iii. Which of the following is a helical virus

a. TMV | b. T4 phage
c. Poxvirus | d. Herpes virus

iv. Which of the following statements are true about a virion

a. Lytic phage
b. Lysogenic phage
c. The viral capsid
d. An infectious and fully formed viral particle

v. A virus is made up of-

a) Protein coat and nucleic acid
b) Protein coat and mitochondria
c) Nucleic acid and cell membrane
d) Nucleic acid, cell wall and cell membrane

vi. Which of the following statements are true about the viruses

a) Free-living | b) Obligate parasites
c) Both (a) and (b) | d) None of these

vii. The shape of the TMV is

a. Rod Shaped | b. Oval Shaped
c. Cuboidal shaped | d. Spherical shaped

viii. The viral genome is enveloped in a protein coat known as

a. Capsid | b. Outer envelope
c. Capsomere | d. Nucleic

ix. The genetic constituent of viruses is

a. RNA | b. DNA
c. ss DNA | d. DNA or RNA

x. A fully formed infectious viral particle is called

a. Virion | b. Viriod
c. Capsid | d. Virusoid

Answer keys

A. Fill in the blanks

i. Living cells **ii.** Lysogeny **iii.** Adsorption **iv.** Genome and Capsid **v.** Roughly Spherical **vi.** UV Light **vii.**Protomers viii.nanometers **ix.** Proteins **x.** D

B. Multiple choice questions

i. **c** ii. **b** iii. **a** iv. **d** v. **a** vi. **b** vii. **a** viii. **a** ix. **d** x. **a**

C. Descriptive type questions

i. Differentiate between lytic and lysogenic cycles.

ii. What are the factors that affect the survival of viruses in soils?

iii. Write the characteristics of bacteriophage.

iv. What are virions, viroids and prions?

v. Write the differences between simple and enveloped viruses?

8

Rhizosphere

To a soil microorganism, the rhizosphere is like a lush oasis in the desert. In comparison to near-starvation conditions of the bulk soil, the rhizosphere is the place where nutrients are plentiful and microorganisms flourish. The term rhizosphere was first used by Hiltner in 1904 to describe the zone of soil under the influence of roots. The rhizosphere can extend more than 5 mm from the root and is the area of increased microbial activity due to exudation of low molecular weight compounds from root cells. The rhizoplane is the surface of plant root and any strongly adhering soil particles. The microbial populations that colonize the interior of the root and from intimate associations with the root are considered endophytes.

Plant roots excrete various organic substances such as amino acids, organic acids, sugars, alkaloids, tannins, phenolic compounds, phytohormones and vitamins. These chemical substances provide a rich source of nutrients for the microbial community and therefore may selectively stimulate the growth of some groups of microorganisms. For example, glucose and amino acids in the exudates readily attract Gram-negative rods, which predominantly colonize roots. Greater populations of microorganisms in the rhizosphere competing for nutrients may create a nutrient deficiency for the plant or microorganisms may make nutrients available to plants from insoluble sources by mineralization process and subsequently increase plant growth e.g. N_2-fixing bacteria, mycorrhizal associations or plant growth promoting rhizobacteria.

Microbial populations of the rhizosphere

A vast number of species of microorganism are present in the rhizosphere and their numbers generally decrease as the distance from the root increases. To measure the effect of the rhizosphere on a particular population, the number of microorganisms in the rhizosphere (R) and the number of microorganisms in the bulk soil (S, soil not influenced by the root) are compared. This R/S ratio provides an estimate of how strongly the rhizosphere affects a particular organism. For example, in wheat rhizosphere the R/S ratio for bacteria, fungi and protozoa was found 24.0, 12.0 and 2.4, respectively. This relationship can also differ with plant species. e.g. the R/S ratio of bacteria for red clover, oats and maize was found 24, 6 and 3, respectively.

Bacteria: These are the most numerous inhabitants of the rhizosphere - typically numbering 10^6-10^9 organisms per gram of rhizosphere soil. they cover 4-10% of the root area occurring profusely in particular microsites. In general, non-sporulating rods are found in great abundance in the rhizosphere. *Pseudomonas* and other Gram-negative bacteria are especially competitive in the rhizosphere and occupy a large portion of the total bacterial population on the root. Many bacteria have a large R/S ratio, indicating marked stimulation in the rhizosphere. The genera most common are *Pseudomonas, Arthrobacter, Agrobacterium, Alcaligenes, Azotobacter, Mycobacterium, Flavobacterium, Cellulomonas* and *Micrococcus*.

Fungi

The plate counts of rhizosphere fungi are generally less than bacteria. Population numbers may best be reported as ranges, most frequently 10 to 20% of the total microflora or 10^5-10^6 g^{-1} rhizosphere soil. The rhizosphere effect, as indicated by R/S ratios, range from 10 to 20. Several rhizosphere-inhabiting fungi are pathogenic, while others form symbiotic (mycorrhizal) associations with roots. The zoospore forming lower fungi such as *Phytophthora, Pythium* and *Aphanomyces* are strongly attracted to roots and can cause diseases when conditions are favourable. Several fungi (*Gibberella fujikuroi*) can produce phytohormones, e.g. Gibberellin hormone and can influence plant growth.

Actinomycetes, Protozoans and Algae

These groups of microorganisms are not significantly influenced by their proximity to roots and the R/S ratio rarely exceeds 2 or 3:1. Actinomycetes account for approximately 10-30% of total microflora in the rhizosphere. The presence of antagonistic actinomycetes may suppress the growth of bacteria. Among the algae, those producing abundant oxygen, may help refurbishing the oxygen level which goes down due to root respiration.

A. Contributions of the plant to the rhizosphere ecosystem

The role of the plant in the rhizosphere is to provide a surface for microcolony development and to produce fixed-carbon compounds used by the soil microbes for carbon and energy, and as growth factors.

(i) Contribution of fixed-carbon compounds or photosynthates to soil microbes

Major questions regarding this contribution process are associated with

- quantifying the quantities of photosynthates entering the rhizosphere
- identifying the organic components of root exudates and

- explaining the variation in these parameters throughout the growing season and between plant species

Because, in native soils, photosynthate produced by higher plants generally provides the primary energy source for the soil microbial community. Therefore, studies are going on for quantifying the proportion of carbon fixed by the plant, released into the rhizosphere community and the ultimate decomposition of fixed carbon (i.e. incorporation into humic substances: 10-20%, microbial biomass: 20-40%, or oxidation to CO_2 by the soil biota: 40-60%).

The rapidity with which carbon derived from the atmosphere can be made available to the underground inhabitants is evident from experiments involving exposure of seedlings to $^{14}CO_2$. Within a short period of 3-4 hrs., ^{14}C may be found in the root exudates.

A useful procedure for assessing the nature and quantity of fixed-carbon products exuded from growing roots would seem to be to quantify the biochemicals accumulated in the growth medium of plants grown under axenic conditions in liquid culture either in a green house or a growth chamber culture. The primary difficulty encountered with such studies is that the microbial community has a stimulatory effect on root exudates production and may alter the nature of the biochemicals contained in root exudates.

Barber and Martin (1976) found that between 5 and 10% of the photosynthetically fixed carbon in wheat and barley plants was lost from the roots grown under axenic conditions, whereas 12 to 18% of the photosynthate was lost from roots growing in unsterilized soils. Using radiolabeled carbon with plants growing in soils appears to provide an improved estimate of photosynthate incorporation into soil fractions. For this, the plants are grown in green house or growth chamber experiments and the photosynthate are labelled with ^{14}C-labelled CO_2. Above ground and below ground portions of the plant must be physically separated so that access of the soil microbial community to $^{14}CO_2$ or 14C-labelled photosynthetically produced organic compounds can ossur only through the root tissue. Movement of the ^{14}C-labelled photosynthate in the rhizosphere is determined by (i) quantifying the 14CO2 evolved from rhizosphere soil, 40-60% (ii) incorporation of 14C into humic substances, 10-20% (e.g. using NaOH fractionation of soil organic matter) and (iii) microbial biomass, 20-40% (e.g. with the chloroform fumigation technique). Nutrients available to rhizosphere microbes vary with plant species, plant development stage, soil temperature as well as the species distribution of the plant community.

Newman (1985) concluded that soluble root exudates ranges from 1 to 10g/100g dry roots. The data reported in literature is inconclusive and some reports of plant carbon incorporated into soil fractions include total plant carbon retained in soil organic fractions, whereas in other studies, root exudate production is evaluated exclusive of carbon contributions by death of sloughed root cells and root death.

Martin (1977) reported a range of rhizodeposition of 14.3 to 44.4% of photosynthetically fixed carbon (for wheat plants at different development stages), whereas Davenport and Thomas (1988) found 10% of 14CO2 fixed by corn (*Zea mays* L.) in belowground plant components as compared to 40% for bromegrass (*Bromus intermis* Leyss). This differential separation of fixed carbon between above ground and below ground plant biomass resulted in a rhizodeposition rate in the bromegrass twice that of the corn.

The composition of root exudates is a primary parameter in selecting for individual species active in the rhizosphere community and the diversity of that community. The composition of root exudates varies with growth condition of the plant and its developmental stage. A variety of compounds that may be contained in root exudates is compiled by Alexander martin (1977). These soluble organic carbon sources accumulated by plants grown aseptically include:

Amino acids	All naturally occurring amino acids
Organic acids	Acetic, butyric, citric, fumaric, glycolic, lactic, malic, oxalic, propionic, succinic, tartaric and valeric acids
Carbohydrates	Arabinose, deoxyribose, fructose, galactose, glucose, maltose, mannose, raffinose, rhamnose, ribose, sucrose and xylose
Purines and Pyrimidines	Adenine, cytidine, guanine and uridine
Vitamins/Growth factors	p-aminobenzoate, biotin, choline, inositol, nicotinic acid, pantothenate, pyridoxine, thiamine
Enzymes	Amylase, invertase, phosphatase, protease
Other compounds	Auxins, glycosides, HCN, p-hydroxybenzoate, peptides, saponin, scopoletin

(ii) The plant assimilates inorganic nutrients, thereby lowing the concentration available for microbial development.

(iii) In respiration of the root, O_2 is consumed and CO_2 is liberated. The utilization of carbonaceous substrates by the large microbial community also leads to the utilization of O2 and release of CO2. Therefore, the respiration of both macro- and microorganisms result in a greater rate of O2 depletion and CO2 production from rhizosphere than from non-rhizosphere soil.

To assess the microbial contribution to the gaseous exchange, the rate of CO2 evolution from sterile and non-sterile roots is measured. Comparison revealed that one-third to two-thirds (33-67%) of the carbon mineralized is the result of microbial respiration.

(iv) Root penetration also improves soil structure and improved structural relationships favour microbial oxidations.

(v) Microorganisms are also affected by root respiration, which alters the pH or the availability of certain inorganic nutrients by the evolution of CO2. The large quantities of CO2 liberated by the rhizosphere inhabitants also influence crop nutrition. By forming carbonic acid, the gas can cause a solubilization of insoluble inorganic nutrients not readily available to the plant. This would effectively increase the supply of assimilable inorganic nutrients. By this means, the level of available phosphorus, potassium, magnesium and calcium may rise.

B. Effect of rhizosphere microorganisms on plants

Just as plant roots have a direct effect on the surrounding microbial populations, microorganisms in the rhizosphere also have a marked influence on the growth of plants. In the absence of appropriate microbial populations in the rhozosphere, plant growth may be impaired. Microbiall populations in the rhizosphere may benefior the plant in a variety of ways, including:

- increased recycling and solubilisation of mineral nutrients,
- synthesis of vitamins, amino acids, auxins and gibberellins which stimulate plant growth and
- antagonism with potential plant pathogens through competitions and development of amensal relationships based on production of antibiotics.

(i) Plants that grow in flooded sediments and soils have evolved adaptations for conducting oxygen from shoot to the roots. but in nearby anaerobic environments, the roots also have to cope with the toxic hydrogen sulfide generated by sulfate reduction. Rice and perhaps other partially submerged plants are protected against the toxic effect of hydrogen sulfide by a mutualistic association with *Beggiatoa.* This microaerophilic, catalase negative, sulfide oxidising filamentous bacterium benefits from the oxygen and catalase enzyme provided by the rice roots and, in turn, Beggiatoa helps the rice plant by oxidising the toxic hydrogen sulfide to harmless sulfur or sulphate, thus protecting the cytochrome system of the rice roots.

(ii) Organisms in the rhizosphere produce organic compounds that affect the proliferation of the plant root system. *Arthrobacter, Pseudomonas, Agrobacterium, Rhizobium/Bradyrhizobium, Azospirillum* and *Azotobacter* populations found in the rhizosphere have been reported to be capable of producing organic chemicals that stimulate growth of plants.

Microorganisms synthesize auxins, cytokinins and gibberellin-like compounds and these compounds increase the rate of seed germination and cause stimulation of root tissue development, thereby increasing the capacity of the root system to provide nutrients and water required for above ground biomass production.

The rhizosphere of wheat seedlings, for example, contains a significant proportion of bacteria that produce indole acetic acid (IAA, a plant growth hormone) that increase the growth of plant roots. In older wheat plants, there are a lower proportion of microorganisms in the rhizosphere capable of producing IAA. This may be in response to a decline in root exudates production, but it also beneficially corresponds to a decreased need for the growth hormone by the plant at this stage.

(iii) A more indirect, but clearly beneficial, result of stimulation of soil community by root invasion is the contribution of rhizosphere microbial communities to development of a stable soil structure conducive to plant community development. The rhizosphere community produces polysaccharide material such as capsules and slimes that cements soil mineral particulates into microaggregates. Furthermore, soil structural improvements are gained from fungal mycelial production as well as by association of soil particles with root tissue. This improvement of soil structure through increased soil aggregation results in improvements in soil aeration,, water infiltration and root penetration.

Fig. The beneficial, harmful and neutral or variable effects of the rhizosphere microbial community on plant growth.

(iv) Allelopathic (antagonistic) substances released by microorganisms in the rhizosphere may allow plants to enter amensal (antagonistic) relationships with other plants. Such allelopathic substances surrounding some plants can prevent invasion of that habitat by other plants, and this may represent a synergistic relationship between a plant and its rhizosphere microbial community.

Bacterial populations in the rhizosphere of young wheat plants have been sown to inhibit the growth of pea and lettuce plants. As wheat plant matures, the proportion of these bacteria decreases and they are replaced by a higher proportion of microorganisms capable of producing growth promoting substances similar to gibberellic acid. Similarly, rhizosphere bacteria (rhizobacteria) may also be used to suppress growth of sensitive plants. For example, Pseudomonas sp. has been isolated from roots of winter wheat (*Triticum aestivum* L.) and downy brome (*Bromus tectorum* L.), which suppress growth of the downy brome weed.

(v) Microorganisms in the rhizosphere influence the availability of mineral nutrients to the plants, sometimes (i) using limiting concentrations of inorganic nutrients before they can reach plant roots, and (ii) in other cases, increasing the availability of inorganic nutrients to the plants. For example, rhizosphere microorganisms increase the availability of phosphate through solubilization of materials that would be otherwise unavailable to plant. Plants have been shown to exhibit higher rates of phosphorus uptake when associated with rhizosphere microorganisms than in sterile soils. The principal mechanism of increasing phosphate availability is the microbial production of acids that dissolve apatite, releasing soluble forms of phosphorus.

Microbially produced organic acids may solubilize essential minerals simply through acidification of rhizosphere soil. The capacity of organic acids excreted by soil bacteria to solubilize soil minerals is exemplified by studies of 2-ketogluconic acid production by Gram-negative bacteria growing in glucose media. Additionally, the acidification of the rhizosphere environments through metabolic production of hydrogen ions alters the pH sufficiently to mobilize soil minerals.

Solubilization of variety of metals contained in natural and synthetic insoluble salts and minerals has been demonstrated. Siderophores and other metal chelators are synthesized by a variety of bacteria, including common soil organisms as well as mycorrhizal fungal associates. Rhizosphere microorganisms that produce chelating agents (e.g. siderophore production) increase the solubility of iron and manganese compounds and thus iron and manganese may be more available to plants. A number of examples of stimulation of plant biomass production by these microbially synthesized metal chelators have been established. rhizosphere is also the most impacted soil microsite for siderophore synthesis.

It has also been shown that microorganisms on roots significantly increase the uptake rates of calcium by the roots. this increase may be due to the high concentrations of CO_2 in the rhizosphere produced by microorganisms, which increases the solubility and thus the availability of calcium. Translocation of various radiolabelled organic compounds and heavy metals along mycelial filaments has also been demonstrated.

Two root-based symbiotic associations are instrumental in producing a thriving, stable plant community: nitrogen fixation associations (*Rhizobium*-legume, *Frankia*-actinorhizal associations) and mycorrhizal symbiosis. The intimate association of the bacterial or actinomycete partner (nitrogen fixation) or fungi with the plant root (mycorrhizae) increases fixed-nitrogen resources or facilitate phosphate nutrient transfer to the higher plants.

(vi) The rhizosphere microflora directly or indirectly inhibit the invasion of the plant tissue by the pathogen. This disease reduction could result from direct competition between the pathogen and root microbes, antagonism of the pathogen by root microbes, or alteration of root exudates diffusion into the soil environment in a manner that interferes with chemotaxis of the pathogen to the root.

(a) Root-inhabiting microbes and plant pathogens may compete for space, nutrients, or even for binding sites on the root surface. Space and nutrient competition could result in failure of the pathogen to develop critical population densities for disease initiation, whereas competition for specific binding sites would reduce the capability of plant pathogen to initiate the infection process.

(b) The antibiotics and siderophores produced by microbial populations in the rhizosphere soil may confer survival advantage. A variety of Pseudomonas strains have been shown to reduce or preclude infection of wheat roots by take all disease due to production of antibiotics or siderophores. Other suggested interactions of antibiotic-producing microbes in biocontrol of root diseases are associated with Rhizoctonia infections and onion white root.

(c) An additional role of rhizosphere microbes in reducing root disease incidence in is interfering with chemotactic attraction of the pathogen to root receptor sites. Chemotactic interactions are more difficult to document in the field but are readily deduced to be of importance of plant disease production. A variety of compounds as components of root exudates may serve as attractants for plant pathogens. Growth of root inhabitants (including mycorrhizal fungi) necessarily reduces both the quantity and diversity of organic compounds diffusing from the root, thereby, diminishing the probability of encounter of a plant pathogen.

(vii) Although increased uptake of minerals due to rhizosphere microorganisms is beneficial. The abundant microbial populations in the rhizosphere can sometimes create a deficiency of required minerals for the plants. For example, the plant disease 'Little leaf' of fruit trees and grey speck of oats are caused by bacterial immobilization of zinc and oxidation of manganese, respectively.

Microorganisms in the rhizosphere may immobilize limiting nitrogen and phosphorus making it unavailable for the plant. Thus, immobilization of nitrogen and phosphorus by microorganisms in the rhizosphere accounts for an appreciable loss of added nitrogen and phosphorus fertilizer intended for plant use. Part of nitrogen is immobilized in the form of microbial protein, but some may also be lost to the atmosphere by denitrification.

Plant microbial interactions in the rhizosphere

Although diverse and complex, the majorities of interactions in the rhizosphere are mutually beneficial to both plants and microorganisms and are synergistic in nature. A further exploration and optimization of these interactions may lead to significant improvements in crop production.

Manipulations of rhizosphere population

Modification of the composition of any soil community structure through inoculation with exogenous microbes requires a thorough knowledge of microbial interactions. Long before the microbiology of the symbiosis was understood, legumes have been exploited in crop rotations for their capacity to increase soil fertility. Biological nitrogen fixation has been encouraged through inoculation of legumes with *Rhizobium/Bradyrhizobium sp.* to enhance symbiotic nitrogen fixation for many decades. This well established soil community modification procedure has been supplemented with methods to supply fungal inocula to stimulate mycorrhizal development or inoculation of a variety of plant-growth stimulating bacteria i.e. *Pseudomonas* and *Bacillus*, including the free-living N_2-fixing bacteria of the genera *Azotobacter* and *Azospirillum*.

Table. Modification of root communities by amendment with alien microbial populations.

Microbe	Recipient	Reference
Rhizobium sp.	Legume seeds	
Frankia sp.	Higher plants	
Mycorrhizal fungi	Higher plants	
Pseudomonas putida	Wheat	Vrany et al., 1981
Pseudomonas fluorescens	Wheat	Parke et al., 1986
	Potato	Bahma et al., 1988
Pseudomonas striata	Wheat and Peas	Astrom and Gerhardson, 1988
Bacillus sp.	Spring wheat	Chanway et al., 1988
Bacillus polymyxa	Lodge pole pine	Holl and Chanway, 1992
Azospirillum sp.	Spring and winter wheat	Bashan, 1986a, b, Harris et al., 1989
	maize	Fallik et al., 1988
Azospirillum brasiliense	Wheat	Bashan et al., 1987

The potential for modifying soil microbial community structure through inoculation with alien strains is frequently successful and only limitations to the procedure are:

- selection of an appropriate microbial strain for inoculation,
- development of a carrier to sustain the microbial population, and
- selection of a delivery system.

Indeed, a variety of strains, carriers and inoculation procedures have been developed but still problem exists, and some inoculation studies are successful while others are not. Some microbial strains are established in the community more readily than others.

A central maxim in soil ecology is that if a niche (functional entity) exists in an ecosystem (a habitat) for a microbe to infection, it will be occupied by a member of the community.

Additionally, displacement of an established, functional member of the community is difficult. Yet, such displacement is frequently desirable in that an improved function of the ecosystem may be achieved by establishing a more efficient or active microbial strain.

Usually, the inoculation of established root systems results in a low probability of development of the inoculated populations. This problem of inoculating established roots has been overcome by drenching (wet all over, soaking) root systems (Parke *et al.*, 1986) or by inoculating seeds before planting. Typically, seeds may be drenched with liquid inocula or pelleted soil carriers of the microbial inoculum prior to planting (inoculation with *Rhizobium* and *Azospirillum sp.*). *Azospirillum brasiliense* is alsomixed throughout the soil and added from plant to plant by growth of wheat roots. Alternatives to root drenches and seed/soil inoculation include banding of the microbial inoculum material into the soil adjacent to the expanding root mass. This procedure is based on the assumptions that the roots will grow through the band of inoculum. The inoculum will survive sufficiently long in the soil to encounter the root and the inoculant will compete successfully with indigenous soil microbes for colonizable surfaces on the root.

Competition of the inoculated microorganisms with indigenous soil microbial populations is the least understood aspect of any soil community inoculation procedure. The added organisms must be capable of overcoming the normal soil community defences to invasion as well as of coping with soil physical and chemical barriers. Many of the growth requirements for the microbe are met by the provision of nutrients and growth surface by the root; hence, dealing with biological competitors becomes the primary difficulty to establishment.

Development of microbial strains that would be useful for field inoculation is generally initiated by isolation of an organism from the indigenous soil population which is amenable for selection of genetic variants (mutants) or by genetic engineering procedures. For example, development of *Rhizobium* strains with improved nitrogen fixation capabilities have been achieved but these efficient strains often fail to enter the site of expression of their enhanced ability - the root nodule. Unfortunately, the laboratory constructed Rhizobium strains are usually the poor competitors for nodule occupancy.

Thus, the rhizosphere microbial community structure has been manipulated through inoculation of the plant with alien organisms to improve plant development (i.e. in most cases, improve crop yields). Great potential exists for further exploitation of these skills to alter the microbial community to accelerate plant community development and to reduce soil pollutant loads. The former objective may be achieved by establishing members of the rhizosphere community with the capacity to decompose plant toxicants or improve nutrient availability or augment soil aggregation, whereas the rhizosphere community may also be used to couple the nutrient generation ability of higher plants (e.g. fixed carbon supplied) with the ability of a laboratory-generated rhizosphere inoculant bacterial species to degrade a troublesome pollutant (perhaps, even cometabolically).

The rhizosphere studies in the field

Most studies of the rhizosphere have been made under laboratory conditions. Some of the early investigations demonstrated that viable counts of specific organisms were greater near roots than in the bulk soil. However, this gives no idea of the biomass (the total amount of living organisms) around roots. To assess biomass microscopically is tedious and recently techniques have become available for indirect measurements of microbial activities such as:

- by assaying ATP of soil.
- by assaying phosphatases and dehydrogenase activities in soil or
- by determining the flush of CO_2 produced from fumigated soil on reinoculation.
- Nitrogenase functioning by determining acetylene reduction activity in soil, denitrification rate measurement and sulphate reduction.

These microbial activities have been found enhanced around roots in soil.

Measurement of numbers, biomass and activities in microbial ecosystems

Understanding the structure and functioning of ecosystems depends upon quantitative information about numbers of organisms, biomass of populations and rates of activity within ecosystems. A change in microbial numbers usually correlates to a similar change in biomass and activity. Microorganisms are extremely diverse and so the methods used to enumerate one group of microorganisms may be inappropriate for the enumeration of another group. Special techniques must be used for enumerating specific physiological groups, such psychrophilic bacteria or obligate anaerobic bacteria. There are two principal approaches used for rnumerating microorganisms; direct microscopic observation and indirect viable count procedure by the plate count technique and most probable number technique.

Biomass is an important ecological parameter because it measures the quantity of energy being stored in a particular segment of the biological community. Biomass literally means 'mass of living material' and can be expressed in units of weight (grams), which can be converted to units of energy (calories). The most practical approach to the determination of microbial biomass is to assay for specific bio-chemicals that indicate the presence of microorganisms. Ideally, all the microbial biomass to be determined should have the same quantity of the biochemical being assayed, so that there is a direct correlation between the amounts of the biochemical measured and the biomass of microorganisms. Also the biochemical being assayed should be present only in the biomass to be determined. The determination of microbial biomass is often carried out by measurement of ATP, cell wall components (e.g. muramic acid, LPS, chitin), chlorophyll and other pigments, DNA concentrations and protein determinations.

Recent advances in analytical methodology provide the sensitivity needed to measure natural rates of microbial metabolic activities within natural ecosystems. One approach to measuring natural rates of heterotrophic microbial activity is to measure uptake rates for tracer levels of radiolabelled substrates to determine the heterotrophic potential for the utilization of that substrate. The rate of DNA synthesis reflects the growth rates of microorganisms and is proportional to biomass. Both heterotrophic and autotrophic assimilation of CO2 can be measured using radiolabelled bicarbonate by incubating a sample containing the indigenous microbial community with the radiolabelled substrate and then determining the amount of 14CO2 assimilated into the cellular organic matter. Moreover, 14CO2 release from labelled substrates can also be used to determine decomposition rates for specific substrates.

The other methods, which do not employ radiolabelled substrates, can also be used to measure rates of microbial respiration. These methods normally measure either rates of oxygen consumption or rate of carbon dioxide production. A variety of enzyme assays can also be used for measuring the metabolic activities of indigenous microorganisms. Some, such as measurement of dehydrogenase and phosphatase activities assay general activities of a relatively large portion of the microbial community; other assays, such as measurement of cellulase, chitinase and nitrogenase activities, assay the metabolic functions of small but important segments of the microbial community. It is important that the microbial community not be altered during the assay procedure if the measurement of enzymatic activity is to reflect *in situ* activities accurately. Caution must be given to maintenance of *in situ* conditions, particularly with reference to temperature, moisture content and E_h and incubation periods must be short enough to preclude changes in number of microorganisms, which could alter the levels of enzymes present in the sample.

Some specific enzyme assays are:

Dehydrogenase	Triphenyl tetrazolium chloride (TTC)	Dehydrogenase converts TTC to triphenyl formazan; the triphenyl formazan is extracted with methanol and quantified spectrophotometrically.
Phosphatase	p-nitrophenol phosphate	Phosphatases convert the p-nitrophenol phosphate to p-nitrophenol which is extracted in aqueous solution and quantified spectrophotometrically.
Protease	Gelatin	Gelatin hydrolysis, as an example of proteolytic activity, can be measured by the determination of residual protein.
Amylase	Starch	The amount of residual starch is quantified spectrophotometrically by the intensity of the blue colour resulting from its reaction with iodine.
Chitinase	Chitin	Production of reducing sugars is measured using anthrone reagent.
Cellulase	Cellulose, carboxymethyl cellulose	Production of reducing sugars is measured using anthrone reagent. Cellulases alter the viscosity of carboxymethyl cellulose which can be measured.
Nitrogenase	Acetylene	Nitrogenase, besides reducing dinitrogen gas to ammonia, is also capable of reducing acetylene to ethylene. The rate of ethylene formation can be monitored using a gas chromatograph with flame ionization detector and the rates of N_2 fixation can be calculated using an appropriate conversion factor.
Nitrate reductase	Nitrate	Dissimilatory nitrate reductase can be assayed by the disappearance of nitrate or by measuring the evolution of denitrification products, such as nitrogen gas and nitrous oxide, from samples using a gas chromatograph. Denitrification can be blocked at the nitrous oxide level by the addition of acetylene, permitting asimpler assay procedure by gas chromatograph.

Factors affecting rhizosphere population

The microflora of the rhizosphere is affected by a number of factors:

1. Proximity of soil sample to the root is particularly important and the bacterial count increases in samples taken progressively closer to the tissue surface. Simultaneously, the total activities of the microbial community, as measured by CO_2 evolution, are enhanced by closeness to the root.
2. The depth of sampling is another important ecological variable and the frequency of bacteria, fungi and algae declines with depth.
3. Different plant species often establish or favour the development of distinct microbial community. The differences are attributed to the variations in (i) rooting habits, (ii) tissue composition and (iii) root exudate products of the host plant. As a rule, legumes have a more pronounced rhizosphere effect than grasses or grain crops. Alfalfa and several clovers have an especially pronounced influence on bacterial population. For example, the R/S ratio of bacteria for red clover, oats and maize was found 24, 6 and 3, respectively. Biennials exerts a more prolonged stimulation than annuals because of their long growth period. At the same time, individual plant species cause a striking response on one or two bacterial genera, for example, *Pseudomonas* and *Bacillus*.
4. The age of the plant also alters the underground microflora. Microorganisms in the rhizosphere undergo successional changes as the plant grows from seed germination to maturity. The successional changes correspond to changes in the materials released by the plant roots to the rhizosphere during plant maturation. In young seedlings, carbohydrate exudates and mucilaginous materials support the growth of large populations of microorganisms on the root surface and within the mucilaginous layers surrounding the roots. As the plant matures, autolysis of some root tissue takes place and simple sugars as well as amino acids are released into the soil, stimulating the growth of *Pseudomonas* and other bacteria with high intrinsic growth rates. When plant stops growing near the end of the growing season and the roots start drying, the readily available carbohydrates are quickly metabolized and the microbial abundance declines.
5. The microbial community of fallow land and non-rhizosphere soil responds greatly to the additions of organic materials but this does not apply to the root zone bacteria. Addition of crop residues, animal manures and chemical fertilizers does not cause appreciable or quantitative

changes in the microflora of root region. Other soil treatments also have little influence on the total number of microorganisms. In general, the character of vegetation is more important than the fertility level of the soil. Different plant species in the same field e.g., chickpea and mustard or bajra and moong or cotton and sorghum have widely divergent number of microorganisms in their rhizosphere while the composition and size of the community under the same species cultivated in fields of greatly differing fertility status fluctuate only to a moderate extent.

6. The kind and yield of root exudate products are governed by the inorganic nutrients availability, temperature, light intensity, O_2 and CO_2 level, root injury and plant age. The release of different organic compounds and their amount affect the growth of different microorganisms in the rhizosphere.

Objective type questions

A. Fill up the blanks with suitable words

i. Dehydrogenase converts TTC to ____________________.

ii. The word Rhizosphere is given by____________________.

iii. The R:S ratio of bacteria is __________________.

iv. The R:S ratio of denitrifier is __________________.

v. The__________is the medial zone directly adjacent to the root including the epidermis and mucilage.

vi. The Greek word "*rhiza*", means__________________.

vii. Acetylene Reduction Assay is used for ________________ enzyme.

viii. microbial populations that colonize the interior of the root and from intimate associations with the root are considered ____________

ix. _____________is the outermost zone which extends from the rhizoplane out into the bulk soil.

x. _____________is the ration of microbial population present in the rhizospheric soil and microbial population present in the non rhizospheric soil.

B. Multiple choice questions

i. The term Rhizosphere was coined by

a. Pasteur b. Lister

c. Hiltner d. Gomes

ii. The medial zone of rhizosphere is known as

a. Endorhizosphere b. Ectorhizosphere

c. Rhizoplane d. Cortical Zone

iii. Exudates come in the form of chemicals released into the rhizosphere by the cells in the roots and cell waste is referred as

a. Rhizodeposits b. Rhizoaccumulates

c. Secretion d. Rhizomix

iv. Compound of low and high molecular weight which are released as result of metabolic process of roots are

a. Secretions b. Lysates

c. Mucilage d. Mucigels

v. Compounds released from autolysis of older epidermal cells of roots are

a. Secretions b. Lysates

c. Mucilage d. Mucigels

vi. Compounds originating in root cap secreted by Golgi bodies of root cap cells are

a. Secretions b. Lysates

c. Mucilage d. Mucigels

vii. Gelatinous material at the surface of root grown in normal non-sterile soil are

a. Secretions b. Lysates

c. Mucilage d. Mucigels

viii. The ration of microbial population present in the rhizospheric soil and microbial population present in the non rhizospheric soil is known as

a. Root effect b. Rhiza effect

c. Microbial effect d. Rhizosphere effect

ix. Which of the following factor/s (is / are) responsible for rhizosphere effect

a. Lysis of roots b. High concentration of CO_2

c. Low conc. of ions d. All of these

x. R:S ratio of ……indicates microbial inhibition

a. Less than 1

b. More than 1

c. Equals to 1

d. None of these

Answer keys

A. Fill in the blanks

i. Tripheyl Formazan **ii.** Hiltner **iii.** 24 **iv.** 1250 **v.** Rhizoplane **vi.** Root **vii.** Nitrogenase **viii.** Endophytes **ix.** Ectorhizosphere **x.** Rhizosphere effect

B. Multiple choice questions:

i. **c** ii. **c** iii. **a** iv. **a** v. **b** vi. **c** vii. **d** viii. **d** ix. **d** x. **a**

C. Descriptive type questions

i. Explain the various sections of rhizosphere.

ii. What are the parameters through which we can assess the rhizosphere effect?

iii. Write the characteristics of bacteriophage.

iv. What do you mean by R: S ratio?

v. Write the various compositions of root exudates?

9

Nitrogen Cycle
Mineralization of Nitrogen

There are two mini-cycles in the soil nitrogen cycle:

(i) Soil-based processes such as mineralization, ammonification, nitrification, and immobilisation include the conversion of organic nitrogenous molecules to plant-useable mineral forms, and subsequently back to biomass (microbial, plant and indirectly animal biomass).

(ii) The process of converting atmospheric nitrogen into forms available to the living components of the ecosystem, i.e. nitrogen fixation and the transformations within the soil, as well as the return of the fixed nitrogen to atmospheric inert forms of nitrogen through denitrification, are all soil and biosphere resident processes.

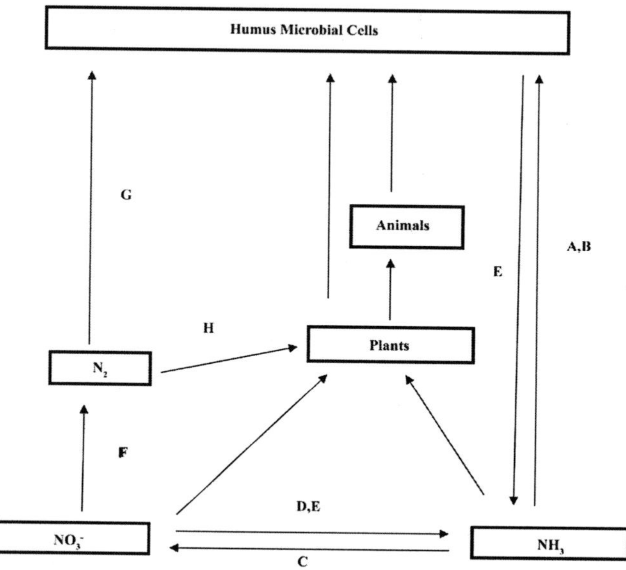

Fig. Nitrogen cycle.

Figure 1. Basic framework of Nitrogen cycle; In this figure, A- Ammonification, B- Mineralisation, C- Nitrification, D- Nitrate reduction, E- Immobilisation, F- Denitrification, G- N_2 fixation (nonsymbiotic), H- N_2 fixation (symbiotic)

- Such transformations are vital for making nitrogen available to plants and their growth. Only a very small part of atmospheric nitrogen is biologically fixed as proteins and other nitrogenous compounds in plants and microorganisms.

Organic nitrogen predominance in the soil ecosystem indicates a potential predictor of plant biomass production capability. Because i) the availability of soil nitrogen resources (along with other potential delimiters of plant productivity such as soil moisture, other nutrient levels, and a variety of soil physical and chemical properties), (ii) the assessment of net mineralization rates of internal soil organic nitrogen pools, and/or (iii) the amendment rates of external nitrogen resources that can provide a valuable estimation of potential producibility. In soils with low amounts of biologically decomposable organic matter or in harvested soils receiving chemical fertiliser or organic residues, nitrogen inputs from sources outside the soil biological community (nitrogen fixation or soil amendment) are the key determinants of ecosystem productivity.A conceptual model of the dependence of plant community productivity and structure on nitrogen mineralization is depicted in figure.

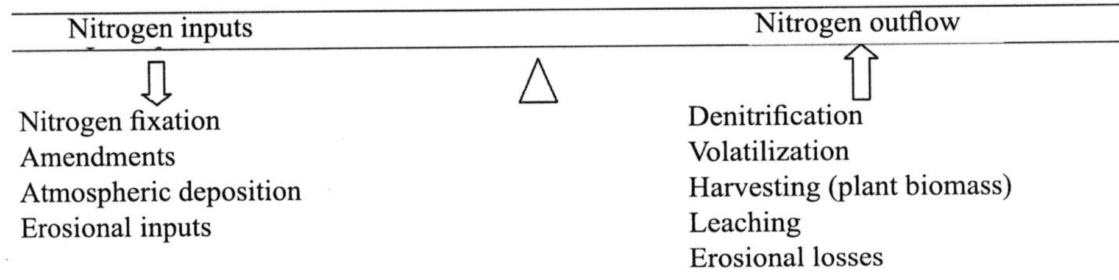

Processes contributing to the soil nitrogen balance

Total biomass production is built on a foundation of nitrogen mineralization. Plant biomass synthesis is controlled by the balance between nitrogen mineralization and immobilisation in an environment with vast reservoirs of organic matter (microbial, plant, and animal biomass, as well as colloidal soil organic matter). Denitrification, volatilization, leaching, soil erosion, and the removal of plant biomass all contribute to the loss of fixed nitrogen (harvesting by animals in native sites and cropping in agricultural systems).These nitrogen outputs must be balanced in a healthy ecosystem by nitrogen fixation, plant biomass inputs, atmospheric deposition, and erosional deposition. A balance occurs between the rates at which mineral nitrogen is created and eliminated in a steady-state system. When these parameters are rebalanced, changing these rates leads in the development of a new biomass productivity steady-state. Crop removal and increased mineral nitrogen leaching are prominent causes of fixed nitrogen depletions in agricultural systems. Excessive organic matter supplements, such as sludges, composts, or animal waste products, can cause system overloads, as can the overuse of nitrogenous fertilisers.

Mineralization of nitrogen

Nitrogen is most common constituent of plants after carbon and oxygen. Nitrogen is almost fully absorbed as nitrate or ammonium in its inorganic state. The majority of nitrogenous materials contained in soil or provided in the form of plant leftovers, on the other hand, are organic and thus mainly unavailable to plants.As a result, understanding its microbial transformation in the soil is crucial. Nitrogen mineralization is the process of microbes converting organic nitrogen to a more transportable, inorganic state. The liberation of CO_2 from carbonaceous materials is akin to this mechanism.

- Both transitions result in the release of inorganic forms of elements.
- Both of these processes are similar in that they are the only way to regenerate the nutrient in a form that can be used for green form development.

Ammonium and nitrate are produced as a result of mineralization, while organic nitrogen is lost. These elements distinguish between two microbiological processes.

Ammonification- is the process of producing ammonium from organic molecules.

Nitrification - in which ammonium is oxidized to nitrate.

Measurements of inorganic (mineral) nitrogen production in humus, proteins, nucleic acids, and related materials are used to determine nitrogen mineralization. The assay of all inorganic products - ammonium, nitrite, and nitrate - is the most appropriate approach. There have been created more advanced procedures that require the stable isotope ^{15}N. Almost all of the nitrogen in surface soil horizons is in the form of organic compounds.

Mineralization is the process by which soil microbes and animals decompose organic molecules into inorganic substances.

Nitrogen mineralization is the process of microorganisms converting organic nitrogen molecules into more transportable inorganic nitrogen compounds such as ammonium and nitrate. Nitrogen mineralization is the formation of inorganic nitrogen, such as ammonium and nitrate, in a broad sense, but it is specifically the production of ammonium, which is why it is also known as ammonification (i.e. ammonium production) or gross nitrogen mineralization.

The connection expresses the net change in the amount of inorganic nitrogen, Ni:

Ni = Nm - (Na+Np+Nl+Nd)

where

Nm is organic nitrogen mineralized

Na is nitrogen assimilated by the microflora

Np is nitrogen removed by the plant

Nl is nitrogen lost by leaching

Nd is nitrogen volatilized by denitrification

The mineralization rate and the velocity of such release in environments receiving nitrogen-rich crop residues ranging from less than 1.0 to 20.0 ppm nitrogen per day are terms used to describe the pace at which organic nitrogen is transformed to ammonium and nitrate. The net amount of inorganic nitrogen produced for a constant amount of metabolised substrate is governed by leaching, denitrification, and, in particular, microbial absorption.

Mineralization of various nitrogen compounds

The nitrogen contained in the soil organic matter occurs in a wide range of compounds from which only a few can be identified. Predominant forms of organic nitrogen in the soil are:

Proteins	20-50%
Amino acids	< 2 ppm
Nucleic acids	< 1%
Amino sugars	5-10%
Vitamins	Traces
Antibiotics	Traces
Metabolic intermediates	Traces
Chitin	Traces
Peptidoglycans etc.	Traces

The organic forms of nitrogen is present in the following macromolecules:

1. Protein: It consists of chains of amino acids, which contains one or more amino sugars. Generally bound amino acids constitute 20-50% of the total humus nitrogen. They are present in the form of proteins or mucopeptides or in the layer polymers with phenolic compounds.

2. Nucleic acids: Usually constitute less than 1%. It consists of purines and pyrimidines bases. A bacterial cell contains 50% proteins, 25% RNA and 3% DNA (on dry weight basis).
3. Chitin is a N-acetyl glucosamine polymer.
4. Peptidoglycan: Polymers in the bacterial cell wall are made up of N-acetyl glucosamine and N-acetyl muramic acid subunits.
5. Amino sugars: It constitute 5-10% in the form of glucosamine and galactosamine.
6. Free amino acids: are found in low quantity, usually less than 2 ppm.

Microbiology

Estimates of population number on various soils reveal that approximately 105 to 107 organisms per gramme of soil are ammonifiers, however these figures are irrelevant because population size is determined by the nitrogen molecule used as a substrate in the growth medium. A simple amino acid, for example, is acted on by a large number of microorganisms, but the nitrogen in chitin is released by only a few taxa.

Proteins provide carbon and nitrogen to a considerable portion of the microflora, and albumin decomposers in the surface horizon range from 105 to 107. Extracellular proteolytic enzymes are produced by microorganisms for the degradation of proteins. The big protein molecule is cleaved into simple pieces by these enzymes. Aerobic bacteria, fungi, and actinomycetes, as well as several facultative and stringent anaerobes, are among the proteolytic organisms.

CO_2, NH_4^+, sulphate, and H_2O are the end products of protein breakdown in aerobic proteolysis.

Anaerobically - When protein-rich materials degrade, foul-smelling chemicals are emitted, a process known as putrefaction. Ammonium, amines, CO_2, organic acids, mercaptans, and hydrogen sulphide are the end products of the anaerobic process.

Pseudomonas, *Bacillus*, *Clostridium*, *Serratia*, and *Micrococcus* species degrade pure proteins quickly. The digestion of casein and the liquefaction of gelatin are key taxonomic features, and proteolysis is a popular criterion in bacteriological classification. Many fungi degrade proteins, amino acids, and other nitrogenous substances quickly, releasing large amounts of ammonium. *Alternaria*, *Aspergillus*, *Mucor*, *Penicillium*, and *Rhizopus* are among the genera having proteolytic enzymes. Fungi, on the other hand, release less NH_4^+

than bacteria because they ingest more nitrogen for cell synthesis. In some soils, particularly in acidic areas, fungi play a significant role in proteolysis.

About 15-70% of the actinomycete isolates from some soils were found to produce protein-hydrolysing enzymes in culture and protein-clearing enzymes are also found among thermophiles of this group. Only old cells that have begun to autolyze release the protein-metabolizing enzymes of certain actinomycetes, and these may in reality be intracellular catalysts liberated on lyses of the hyphae. The actinomycetes' sluggish growth patterns, on the other hand, make comprehending their role in soil proteolysis challenging.

Mineralization process:

The conversion of organic nitrogen molecules to ammonium is mediated by many enzymes generated by bacteria and soil animals. Extracellular enzymes break down organic nitrogen polymers and convert them to monomers, which subsequently pass through the cell membrane and are digested, producing ammonia.

Enzymes involved in mineralization

A. Extracellular enzymes

Protein, aminopolysaccharides, nucleic acid, and urea, among other things, are depolymerized by the principal extracellular enzymes generated by microorganisms.

1. Proteins: Proteinases, commonly known as proteases and peptidases, are enzymes that break down proteins. Peptidases cleave dipeptides or monopeptides, whereas proteinases function on longer protein molecules.These enzymes are substrate specific but hydrolytically cleave the peptide linkage and give rise to individual amino acids. Examples: subtilisin, clostripain and thermolysis are isolated from soil microbes. The resulting amino acid can be used as a carbon, energy, or nitrogen source, or it can be used to make new proteins.
2. Nucleic acid: Ribonucleases (RNases) and deoxyribonucleases (DNases) breakdown nucleic acid by hydrolyzing the ester link between the phosphate group and the pentose sugar. Exonucleases, which split off a single nucleotide from the end of a nucleic acid polymer, and endonucleases, which cleave within the nucleic acid polymer, are the two types of RNases and DNases now known.
3. Chitin: Chitin is a N-acetylglucosamine polymer. Chitin is present in the exoskeletons of insects and forms the cell wall of fungi. Although

microbial cell walls are assumed to be relatively refractory in soil, extracellular enzymes can destroy them. It can be destroyed by chitinase and chitobiase working together.Chitinase breaks it down into chitobiose (dimers), which are then cleaved into two molecules of glucosamine by chitobiase.

4. Peptidoglycan: This is present in the bacterial cell wall. Lysozyme plays important role in its degradation. It disrupts the -1,4 connection between N-acetylglucosamine and N-acetylmuramic acid.

5. Urea: The decomposition product of nucleotides is urea. Urea is also present in animal excreta. Urea is one of the fastest growing source of solid fertilizer 'N' used in agriculture. It is mineralized by enzyme urease that is commonly found in soil bacteria and as an extracellular enzyme in soil. Urease hydrolyses urea into CO_2 and ammonia. Nickel is the cofactor associated with the active sites of some of the ureases. Urea-degrading bacteria are about 32-70% of the total bacterial population in soil and urea-degrading fungi forms about 58-100% of total fungal population. Bacteria, fungi, and actinomycetes produce urease and hence use urea as a nitrogen and energy source. *Bacillus*, *Micrococcus*, *Pseudomonas*, *Klebsiella*, *Corynebacterium* and *Clostridium* are among examples. By enriching soil with a high concentration of urea, urea bacteria can be easily separated from manure or soil.

During mineralization of urea by urease, H+ is consumed. As ammonium is created, the pH of the soil receiving urea rises.If the pH became 9.0, then product formed will be ammonia and not ammonium. Hence a large amount of organic nitrogen will be lost in the atmosphere as gaseous ammonia. Increasing temperature also can increase the loss of urea.

B. Intracellular enzymes

The majority of the time, ammonium is produced within the microbial cells by internal enzymes.

1. Amino acids have two types of nitrogen functional groups: amine (NH_2-R_3) and amide (NH_2-CR=O). Deamination is the mechanism by which amino acid dehydrogenase and amino acid oxidases liberate amino nitrogen. Asparaginase and glutaminase cleave the amide group of asparagine and glutamine, respectively.

2. Amino sugars are broken down into two parts. As previously explained, amino sugar is phosphorylated by kinase, and subsequently ammonia is produced by the deamination process.

3. Nucleotides are broken down into nucleosides and phosphates before being hydrolyzed. Purine, pyrimidines, and pentose sugar are formed once the nucleosides are hydrolyzed. Ammonium is released during purine and pyrimidine metabolism, creating urea as an intermediary.

Because microbial degradation of amino acids, amino sugars, and nucleic acid is an essential need of heterotrophic microorganisms for energy and carbon, the ammonium generated as a result of ammonification might be regarded a by-product of catabolism.

Slow mineralization rate

Mineralization rate refers to the rate at which organic nitrogen is transformed to ammonium and nitrate. The nitrogenous chemicals found in soil organic matter have a long shelf life. Only a small amount of the nitrogen reservoir is mined each growing season because it is so resistant to attack.Several hypotheses are given to account for the slow mineralization.

One hypothesis states that nitrogenous compounds form complexes with the phenols or polyphenols and hence the nitrogenous compounds are less susceptible to digestion. experiments have demonstrated that (i) proteins complexed with polyphenols and (ii) amino acids linked with phenols into polymers are resistant to microbial attack, in contrast with the free proteins and amino acids. Clay minerals, according to a second theory, safeguard nitrogenous chemicals by confining them within the lattice of clay crystals. Extracellular proteolytic enzymes, which are proteins, are adsorbed by clays and may become less active as a result.

Organic N and organic C mineralization are connected to one another. The two elements are mineralized at similar rates in unamended soil, and the ratio of CO_2-C to Ni generated is virtually constant at 7 to 15:1. Soil that is active in one transformation is likely to be active in another. As the rate of Ni production increases, the ratio of C mineralized to N mineralized drops. As a result, microflora that is actively producing nitrate tends to emit C and N in a 7:1 ratio (vigorous). Those who are the least active have ratios of around 15:1. (less active). The introduction of exogenous substrates can alter the equilibrium; for example, protein-poor residues favour CO_2 evolution whereas protein-rich materials favour Ni release. Because water soluble nitrogen is commonly the percentage most rapidly converted to ammonium in plant materials, the speed with which ammonium is released from crop residues are frequently related to the amount of water soluble nitrogen present. Many nitrogenous compounds are found in the soil, but little is known about the immediate source of ammonium generated during their decomposition.

Fate of ammonium in soil

Ammonium has several fates in soil, in addition to the mineralization/ immobilization cycle:

- It can be chemically held on cation exchange sites or become fixed in the lattice of clay minerals (ammonium fixation), such as illite and vermiculite.
- Ammonium can react chemically with organic compounds such as quinones or be volatilized at high pH.

Because ammonium ions can be attached to negatively charged clay particles and are not leached out of the soil, they are chemically stable in acidic circumstances. In acidic soils, on the other hand, nitrate and nitrite (negatively charged ions) are easily leached out, with nitrite also being transformed to gaseous nitrogen and nitrous oxide. However, one downside is that only a limited percentage of ammonium ions attached to clay particles can be exchanged and used for absorption. They may be hazardous to plants at certain concentrations, and they may be volatilized in alkaline soils.

Major biological fates are

- Plant uptake of ammonium nitrogen: Plants assimilate ammonium nitrogen in many forest, orchard, and grassland soils when nitrifying bacteria are inactive due to soil and climatic circumstances. Otherwise, plant roots preferentially absorb nitrates over ammonium salts, indicating that nitrification is important for soil fertility
- Microbial absorption or nitrification microorganisms' oxidation to nitrate

Net ammonium production

There are many factors responsible for net production of ammonium by microorganisms in soil. If nitrogen is easily available then net ammonium production occurs. Let us say if we add a residue of organic compound that is rich in nitrogen such as manure, then it will lead to the production of more inorganic nitrogen. Net ammonium generation happens when residues with a C:N ratio less than 20:1 are added to soil, according to decades of research. The crucial C:N ratio can be estimated using fundamental concepts based on the C:N ratio of heterotrophic bacteria and the yield coefficient, which is the quantity of carbon in the substrate converted to microbial biomass.

Bacteria have a C:N ratio of 4 or 5:1, while fungi have a C:N ratio of 10 or 15:1.

The type of organic matter, microbes, and ambient conditions all influence the yield coefficient.

1. Organic matter: Organic molecules that are easily decomposable, such as simple sugar, have a high yield coefficient of 0.6, whereas lignin has a yield value of less than 0.1.
2. Microorganisms: Fungi are typically more efficient in biomass production, so yield coefficient is 0.5 (0.44) while bacteria have yield coefficient 0.4 (0.32).
3. Environmental conditions: When conditions are unfavourable for microorganisms, the yield coefficient drops because cell maintenance consumes more energy than growth. Also, because anaerobic metabolism produces less energy, the yield coefficient is lower under anaerobic conditions.

Biological factors have an impact on net ammonium production. Soil animals that act as predators for primary decomposers play the most crucial role. Soil animals, such as protozoa and nematodes, are responsible for around 30% of net nitrogen mineralization. Because predators have a C:N ratio comparable to their prey, ammonium is frequently produced as a waste product when they pray on microbes, resulting in an excess of nitrogen due to the loss of carbon as CO_2 during metabolism.

Calculation of C:N ratio

Fungi typically make up about 2/3 of the total soil biomass while bacteria constitute 1/3 of the total biomass.

Also fungi convert 44% of substrate c into cell biomass. So yield coefficient (Y) is 0.44%, similarly bacteria have Y equal to 0.32.

Fungi cell commonly have C:N ratio = 10:1

Bacterial cell commonly have C:N ratio = 4:1

Step 1 - To calculate average yield coefficient and substrate C:N ratio

Average yield coefficient (Y) = 2/3 (0.44) + 1/3 (0.32) = 0.4

Average C:N = 2/3 (10) + 1/3 (4) = 8

Step 2 - To calculate how much microbial biomass is produced

100 g substrate C added = 60 g CO_2-C + 40 g biomass C of microorganisms

Step 3 - To calculate how much microbial biomass N is produced

40 g microbial biomass C/C:N ratio (average C:N = 8)

= 5 g microbial biomass N

Step 4 - Calculate substrate C:N ratio that would be needed

Substrate C/N ratio = 100 g substrate C / 5 g substrate N

= 20:1.

Environmental influences

The metabolic variability of the microflora that causes nitrogen mineralization is an important component in identifying how environmental conditions influence the transformation. Because many type of microorganisms function in degradation of nitrogenous material like -

- Aerobic and anaerobic
- Acid sensitive and acid resistant
- Spore forming and non-spore forming etc.

As long as microbial multiplication is possible, at least some portion of the population is active, regardless of the habitat's popularity. The rate of mineralization, on the other hand, is greatly influenced by the environment. The rapidity of mineralization is governed by physical and chemical features of the habitat, such as moisture, pH, aeration, temperature, and inorganic nutrient availability. It also depends on factors such as vegetation, geography, and age.

(i) ***Total N content of soil:*** The rate of inorganic N production is proportional to the total N content of the soil. In a given time interval, N-rich sites liberate more inorganic ions than N-deficient sites. Because of the relationship between Ni production and total N, the amount freed throughout a growth season under ideal climatic circumstances may be estimated. In the North-Eastern United States, for example, summer crops make 2 to 4 percent of total humus N available to them, whereas winter crops make only 2 percent or less humus N available during the peak growth period. As a general guideline, 1 to 4% of soil organic N released to plants during the growing season in temperate latitudes can be used to estimate the amount of soil organic N released to plants during the growth season.

(ii) ***Nature of organic N compound:*** Mineralization from different sources is ranked as:
Urea>Amino acid>Proteins>Nucleic acids>Amino sugars>Humified nitrogen

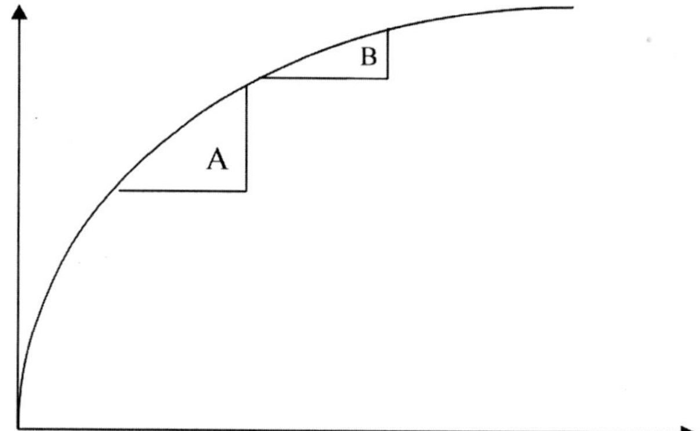

It is observed that for some amount of time less NH_4^+ is formed in region B then in A. It is because neither all the types of organic soil N are equally available nor do they mineralized at the same rate. Rate of mineralization slowed down in region B because the most available N substrates are mineralized first with faster rate; then less available nitrogen substrate which are mineralized with slower rate.

(iii) Moisture: At low water levels, ammonium is generated slowly, but as the moisture level rises, mineralization accelerates. The onset of rainfall is often associated with a rapid initiation of mineralization in desert or semiarid regions, as well as in climatic zones with wet-dry cycles. Although the specific moisture values that are ideal for the conversion fluctuate amongst soils.

(a) The ideal moisture for ammonification is between 50 and 75 percent of the soil's water holding capacity.

(b) Soil submergence does not stop ammonification, and it happens quickly in wet paddy fields with low oxygen levels. In general, soils that are active in aerobic mineralization form ammonium fast in the absence of oxygen, whereas soils that are slow to mineralize in aerated circumstances produce ammonium slowly in anaerobiosis. However, the amount of O2 has an impact on the final inorganic product, as ammonium is only transformed to nitrate in well-aerated habitats.

(c) Drying and wetting cycles speed up mineralization compared to keeping them moist all the time, while the rate of mineralization slows down as the cycles progress. The drying and wetting cycle may make previously inaccessible substrates more accessible, or it may promote cell disintegration. However, none of the proposed hypotheses for this unusual microbial evolution have persuasive data to back them up.

(iv) ***pH:*** The pH of the environment has an impact on mineralization. In neutral settings, the generation of inorganic nitrogen (ammonium plus nitrate) is greater than in acidic situations. Although pH has no effect on transformation in some soils. Mineralization is slowed, but not stopped, by acidification. Liming acid soils has a stimulatory effect because the pH is brought closer to the ideal for active microorganisms. Organic N accumulates in very acidic soils due to sluggish mineralization, resulting in a quick release when these soils are limed.

(v) ***Temperature*****:** As a temperature-sensitive enzyme produced by microorganisms whose growth is conditioned by temperature catalysed each biochemical step, temperature has an impact on the mineralization sequence. The microflora progressively mineralizes the organic compounds at 20°C. The mineralization of nitrogen increases in proportion to the increase in temperature from the freezing point to the optimum. In contrast to most microbiological transformations, the optimum temperature for ammonification is above 40°C, usually between 40 and 60°C, exhibiting the activity of thermophiles and, as a result, ammonium accumulates in composts and manure piles maintained at 65°C.

When compared to soils not exposed to freezing, freezing followed by thawing, like drying followed by wetting, often enhances the rate of humus nitrogen breakdown; the cause for this impact, like the drying-wetting cycle, is unknown.

(vi) ***Addition of inorganic nitrogen:*** The addition of inorganic nitrogen can sometimes improve humus mineralization, as seen by Ni build up or plant uptake of the element. By labelling the added inorganic nitrogen with 15N, the enhancement could be distinguished to that formed from soil constituents. On the other hand, such amendments appear to have no influence in other soils; the stimulation, where it occurs, may result from the response of the community to the addition. The larger and more active community being able to degrade the nitrogenous complexes more readily than the original microflora.

(vii) ***Cultivation:*** Cultivation of certain virgin (uncultivated) lands causes a massive decline in their content of organic nitrogen. This promotion of mineralization is associated with and probably is dependent upon the parallel decline in organic carbon content of virgin lands that are brought into food production. So more nitrogen is released as ammonium, which is then oxidised to nitrate. That causes nitrate pollution of underlying waters, and ultimately pollution of adjacent bodies of surface water may occur.

Significance of mineralization

(i) Since N is essential plant nutrition, considerable efforts have been devoted to devising a technique for predicting the quantity of soil N that will be made available to a crop during a given growing season. If this quantity were known, then the amount of fertilizer that would have to be applied could be predicted and excessive rates of fertilization could be avoided.

The problem of establishing a suitable technique is essentially one of devising a means of predicting the microbial activity in the field by some suitable operation in the laboratory. Two general approaches have been employed.

a. Determining the quantity of N mineralized in a reasonably short incubation period in the laboratory.
b. Establishing what fraction of humus N is correlated with N mineralization rate.

The latter (b) approach is made difficult because the chemical identity of soil constituents that are mineralized in a growing season is unknown, and hence arbitrary fractionation procedures have been introduced in the hope that one or more would give substances whose yield would correlate with extent of mineralization in a designated period. The vast effort to devise such a test for nitrogen availability has led to two sorts of results.

For some soils and in certain regions or with particular crops, significant correlations have been obtained between nitrogen uptake by plants and either mineralization rate or the yield of nitrogen by some extraction technique. A suitable laboratory test would thus seem to have been found. On the other hand, with other soils, in other regions or with different crops, no correlation was evident and hence an acceptable laboratory procedure would appear to be still lacking. Some investigators have proposed the use of a fungus, often an *Aspergillus* strain, to measure the quantity of nitrogen made available by the community. But in this procedure also, the results have not been entirely satisfying.

For the present, therefore, it appears that incubation or extraction procedures reliably predict the nitrogen made available to higher plants in some situations, but for reasons not as yet clear, the very same methods have little predictive value under other circumstances.

(ii) Mineralization has also been the focus of considerable recent interest owing to the use of land for the disposal of large quantities of organic materials, either as digested sewage sludge from urban regions or as animal manure from feedlots containing thousands of cattle. If mineralized, the organic nitrogen could be released as ammonium that could then be oxidised to nitrate. Nitrate entering the groundwater in high concentrations is undesirable from the standpoint of public health. In addition, when large quantities of the wastes are spread in the field, free ammonia may be released in large amounts to the atmosphere and on occasion, presumably when manure is present in volumes such that aeration is inadequate to maintain aerobic populations, various volatile amines may be evolved as well.

The public concern with environmental quality has led to assessments of the formation of many volatile compounds and these evaluations have led to estimates that mineralization gives rise to almost $5x10^{12}$ kg of atmospheric ammonia - N per year, a figure about eight times greater than that associated with the production of all the nitrogen oxides. Even in over-grazed pastures, a considerable quantity of ammonia and possibly volatile amines may be discharged to the atmosphere.

In some instances, thousands of kilograms of nitrogen may be mineralized per hectare in a few years. After a number of years, the decline ceases and the level of humus nitrogen remains at a new steady-state concentration. The protective properties of nitrogenous chemicals in clay soils, as well as lignins and polyphenols, are also significant. Because the resistance to attack on nitrogenous molecules in the soil organic fraction is so high, just a small amount of the soil's nitrogen store is mineralized each growing season.

Objective type questions

A. Fill up the blanks with suitable words

i. ________________ is the process by which soil microbes and animals decompose organic molecules into inorganic substances.

ii. Chitin is a polymer of ____________________

iii. Protein is a chain of ____________________

iv. Enzymes responsible for degradation are____________ and ____________

v. The decomposition product of nucleotides is____________

vi. Bacteria have a C:N ratio of ____________________

vii. Fungi have a C:N ratio of ________________

viii. The oxidation number of dinitrogen is ________________

ix. The ideal moisture for ammonification is between ________________ of the soil's water holding capacity.

x. The microflora progressively mineralizes the organic compounds at ________________

B. Multiple choice questions

i. Nitrogen fixation is the conversion of

a. N_2 to N
b. N_2 to NH_3
c. N_2 to NO_3^-
d. N_2 to Urea

ii. Mineralization is a ....step process

a. Two
b. Four
c. Six
d. Eight

iii. Who scientifically formulated the concept of N cycle

a. Beijerinck
b. Winogradsky
c. Lohnis
d. Needham

iv. Which of the following is true in terms of decomposition

a. Urea>Amino acid>Proteins
b. Amino Acids>Urea>Proteins
c. Proteins>Urea>Amino Acids
d. None of these

v. The optimum temperature for ammonification is usually between

a. 10 to 20°C
b. 20 to 30°C
c. 20 to 25°C
d. 40 to 60°C

vi. How many molecules of ATP are required to fix one molecule of nitrogen

a. 12
b. 20
c. 6
d. 16

vii. Important enzymes involved in nitrogen fixation are

a. Nitrogenases and Hydrogenases
b. Nitrogenases and Hexokinases
c. Nitrogenases and Peptidases
d. Nitrogenases and Hydrolases

viii. The Habers process refers to which biogeochemical cycle

a. S cycle
b. N cycle
c. P cycle
d. Water cycle

ix. How much of the atmosphere is composed of Nitrogen

a. 87%
b. 37%
c. 78%
d. 21%

x. The oxidation number of N in NH_3 is

a. +3
b. -3
c. 0
d. -1

Answer keys

A. Fill in the blanks

i. Mineralization **ii.** N-acetyl glucosamine **iii.** Amino acids **iv.** Peptidases and Proteases **v.** Urea **vi.** 4-5:1 **vii.** 10-15:1 **viii.** Zero **ix.** 50-75% **x.** 20°C

B. Multiple choice questions

i. **b** ii. **a** iii. **c** iv. **a** v. **d** vi. **d** vii. **a** viii. **b** ix. **c** x. **b**

C. Descriptive type questions

i. Write the differences between nitrification and denitrification.

ii. Briefly sketch nitrogen cycle showing its important components?

iii. What do you mean by mineralization and immobilization in soils?

iv. What are the fates of ammonium ions in soils?

v. Write the factors that affects N mineralization in soils?

10

Nitrification

At the point where ammonium is produced, the reactions involved in organic nitrogen mineralization come to a halt. This most reduced form of inorganic nitrogen is the starting point for the nitrification process, which involves the biological synthesis of nitrate or nitrite from reduced nitrogen molecules. The ability of nitrifying bacteria to create nitrate, which is the primary nitrogen source ingested by higher plants, determines their relevance. Nitrate is formed in a variety of places, including soil, marine environments, manure piles, and sewage treatment, where it is the result of the final stage of turning organic nitrogen inoffensive.

During the early nineteenth century, the phenomena of nitrate synthesis advanced, and the product created was thought to be of a chemical nature, formed by the reaction of air oxygen with ammonia, with the soil acting as a chemical catalyst. Pasteur, on the other hand, believed that nitrate creation was microbiological and akin to the conversion of alcohol to vinegar. The earliest experimental proof that nitrification is a biological process was provided by Schloesing and Muntz (1877). They mixed sewage effluent with sterile sand and $CaCO_3$ in a long tube. The liquid's ammonium concentration stayed unchanged for 20 days, but suddenly ammonia was destroyed and nitrate developed.The transition was stopped by heating the column or adding an antiseptic, but it could be restarted by adding a small amount of garden soil.

S. Winogradsky's pioneering study (1890) identified the relevant agents and proved that nitrification is usually related with the metabolism of chemoautotrophic bacteria. In nitrification, two discrete and distinct stages can be distinguished under certain conditions. Because nitrite is produced often during ammonium oxidation, early microbiologists assumed that the transition involved an initial oxidation to nitrite and then a conversion of nitrite to nitrate. When two groups of bacteria were isolated and reported, each catalysing a different piece of the chemical sequence, the correctness of this theory was established.

$$2NH_4^+ + 3O_2 \longrightarrow 2NO_2^- + 4H^+ + 2H_2O$$

$$2NO_2^- + O_2 \longrightarrow 2NO_3^-$$

Nitrifying bacteria

The pioneering work of S. Winogradsky established that nitrification is typically associated with the metabolism of certain chemoautotrophic bacteria. Two groups were distinguished: one deriving its energy for cell synthesis by theoxidation of ammonium and the other group by oxidation of nitrite. The nitrifying autotrophs form no endospores and differ in shape to include - rods, ellipsoids, cocci and spirilla. Despite their morphological heterogeneity, these nitrifying bacteria show physiological similarity with respect to the energy-yielding reactions, the energy for the growth is derived solely from the metabolism of ammonium to nitrite and nitrate. The following genera are known to occur in soil.

(i) Chemoautotrophic bacteria which oxidize ammonium to nitrite

- *Nitrosomonas*: ellipsoidal or short rods
- *Nitrosococcus*: spherical cells
- *Nitrosospira*: spiral shaped cells
- *Nitrosolobus*: lobate and pleiomorphic cells
- *Nitrosovibrio*: curved slender rods

(ii) Chemoautotrophic bacteria which oxidize nitrite to nitrate

- *Nitrobacter*: short rods
- *Nitrococcus*: Cocci motile
- *Nitrospira*: Spiral or vibroid cells
- *Nitrospina*: Slender rods, halophilic marine

Nitrosomonas and Nitrobacter are the only ammonium and nitrite oxidising organisms that are often observed. These are without a doubt the most important nitrifying chemoautotrophs. Nitrosococcus species are uncommon in nitrifying enrichments and have only been recorded in a few cases. Nitrosospira and Nitrosolobus, the other two genera, are infrequently found and do not appear to be abundant in nature.

The nitrifying autotrophs rely on inorganic sources for energy, which makes them obligatory. The carbon for cell synthesis comes entirely or partially from CO_2, carbonates, or bicarbonates, while the energy for CO_2 reduction comes from inorganic nitrogen molecule oxidation. In normal laboratory media, growth does not occur, and many of the medium constituents are even bactericidal.Nitrosomonas and Nitrobacter cells are able to assimilate and metabolize certain simple organic molecule. Although it is unlikely that such assimilation is of much value for their growth in nature. These bacteria don't need any growth factors, and they can make all of the polysaccharides, structural ingredients, amino acids, and vitamins found in their protoplasm from inorganic nutrients and CO_2. These suggests to the bacteria's metabolic complexity as well as their extraordinary nutritional simplicity. All enzymes and other components required for life can be synthesised by bacteria from inorganic elements.

A specific quantity of energy is required in nitrifying bacteria to reduce CO_2 to the type of carbon compounds found in microbial protoplasm. Because inorganic oxidation is the driving force for reduction, the biochemical efficiency of these organisms can be easily represented as the ratio of inorganic nitrogen oxidised to CO_2-carbon absorbed, or a N:C ratio. The ratio of ammonium-nitrogen oxidised to CO_2-carbon absorbed varies amongst *Nitrosomonas* strains, ranging from 14 to 70:1. The ratio of nitrite-nitrogen oxidised to CO_2-carbon fixed in *Nitrobacter* ranges from 76 to 135:1. The chemoautotrophic bacteria of the first phase of nitrification have the following reaction:

$$2NH_4^+ + 3O_2 \longrightarrow 2NO_2^- + 4H^+ + 2H_2O$$

The observed disappearance of substrate, accumulation of product, and use of O2 correspond with the theoretical values for the equations.

Microorganisms do not use all of the energy available to them; in fact, just a small part of it is used. The proportion is decided by their energy use efficiency. The free energy efficiency of *Nitrosomonas* is typically 5-14 percent, while *Nitrobacter's* efficiency is estimated to be 5-10 percent. In younger cultures still in the logarithmic phase, the free energy efficiency is higher than in older cultures.The ammonium oxidizers have lower N:C ratios than the nitrite oxidizers, indicating that the former *Nitrosomonas* obtain more energy from their autotrophic process, requiring less nitrogen to be turned over every cell created.

Occurrence

The quantity of these autotrophs in nature has been studied by several microbiologists. The number of ammonium oxidizers per gram of soil has been observed to range from zero to one million or more. Larger populations are seen in soil with a pH greater than 6.0, however many neutral and alkaline sites contain modest populations. *Nitrosomonas* and *Nitrobacter* species are found together in most ecosystems; otherwise, nitrite levels could reach phytotoxic levels. The usage of ammonium salts has been shown to increase the population of both groups, with values exceeding 10^7 per gram being reported.

Ammonium oxidation can also happen in situations where there are a lot of soluble carbonaceous compounds, including sewage and manure. As a result, in nature, organic stuff does not prevent ammonium oxidation. The well-known impact of carbohydrates in lowering soil nitrate levels is due to a depletion of inorganic nitrogen supply by flora that requires inorganic resources for carbohydrate degradation, rather than an effect on nitrate-producing bacteria (inorganic nitrogen is immobilized more along with more carbohydrate availability and nitrification is not inhibited).

In most cases, nitrite does not accumulate in soil, and nitrate is the most common nitrogen-containing anion. Because of its toxicity to plants and microorganisms, the presence and persistence of nitrite could have a significant impact on crop production. In alkaline soils, for example, nitrate generation from ammonium sulphate is reduced as the rate of ammonium application is increased, although nitrite may accumulate in some circumstances. The nitrite generated is not further converted to nitrate, rather than ammonium oxidation being inhibited. So that it lasts as long as there is ammonium available.

Field studies indicate that nitrite accumulation is the result of two factors.

- Alkalinity
- High ammonium levels

a. In calcareous soils, nitrite accumulation is proportional to the rate of ammonium addition, and the effect rises with decreasing hydrogen ion concentration, leading to alkalinity levels for constant nitrogen fertilisation.

b. Anhydrous ammonia, a common fertilizer, may cause the pH to rise, sometimes to values of 9.0 and 9.5 and the nitrogen source together with the locally high pH frequently causes nitrite accumulations at the site of application.

c. An equivalent mechanism can be found in the breakdown of proteinaceous materials or urea, where the release of ammonium can result in a large but usually transient inhibition of nitrate production.

As the pH or ammonium level falls with progress of nitrification, the suppression is relieved and nitrate production commences. As a result, poisoning from nitrite build-up appears to be an issue on occasion. The build-up of nitrite is attributable to the *Nitrobacter* group's high sensitivity to ammonium salts during the alkaline reaction. Nitrobacter'senergy-producing process is suppressed by as little as 1.4 ppm ammonium-nitrogen at pH 9.5, although the ammonium oxidizers are unaffected. Such observations suggest that ammonium, the natural substrate for nitrification, is a selected inhibitor in alkaline environments of the second step in nitrification by virtue of its toxicity to *Nitrobacter.*

The active principle appears to be free ammonia, which is preferred by pH values higher than 7.0. The nitrite oxidizers develop quickly and destroy their substrate whenever the ammonia level falls to a low level due to oxidation or a reduction in pH during nitrification.

Heterotrophic nitrification

Enrichment procedures are the mechanism by which autotrophs are isolated, and they are designed to be particular for definite physiological kinds. As a result, nitrification is frequently assumed to be entirely chemoautotrophic. Heterotrophicnitrification is currently recognised as a widespread microbiological process, at least in vitro.

a. A large number of heterotrophic bacteria and actinomycetes are able to generate traces of nitrite when grown in culture media, containing ammonium salts. As a rule, nitrite appears only after (i) active growth has largely ceased and (ii) in media in which the ammonium supply far exceeds that required for assimilatory purposes; that is, where the C:N ratio is low.

   Many organisms do not form nitrate. The trace quantities of nitrogen oxidized are in contrast to the 2000 ppm nitrogen or more transformed by *Nitrosomonas*.

b. Several fungi are also capable of oxidizing nitrite in culture, so that two heterotrophic populations acting together can presumably convert ammonium to nitrate. In medium containing ammonium as the sole nitrogen source, a few bacteria, such as *Arthrobacter* strains, and fungus, such as *Aspergillus flavus* and related aspergilli, may create nitrate. The heterotrophs do not make use of the energy released in the oxidation for growth and indeed, often the more oxidized products do not appear until active proliferation has ceased.

Many microbial metabolites contain nitrogen in a partially or highly oxidised state. These include the following microbial metabolites with oxidised nitrogen:

a. N oxides: aspergillic acid, pulcherriminic acid
b. N - hydroxy compounds: hedacidin
c. Hydroxamates: sideramines, mycelianamide

d. Oximes: oximinopropionic acid

e. C - nitroso compounds: alanosine, 1-nitrosoethanol

f. Nitrosoamines

g. Nitrosoamides: streptozotocin

h. C -nitro compounds: 3 - nitropropionic acid, chloramphenicol and others

i. Nitramines: fragrin, nitroaminoacetic acid, β-nitroaminoalanine

The precursors of such metabolites in culture are not always ammonium or amino compounds but the potential of many heterotrophs for the formation of such an array of substances - by oxidative or reductive sequences adds to new dimensions to explain which organisms oxidize nitrogen compounds in nature. The reaction need not stop at the organic product, moreover, in as much as nitro compounds, oximes and undoubtedly others from among these types of chemicals can be transformed to nitrite and sometimes nitrate.

It's difficult to gauge the importance of heterotrophic nitrification. Heterotrophs are not currently implicated in most cases of nitrite or nitrate generation in soil, but they may be present in specific ecosystems.For example,

i. Certain soils contain few autotrophs and yet nitrate is synthesized and,

ii. Sometimes nitrate is formed at temperatures too high for significant replication of the chemoautotrophs.

iii. Moreover, the results of some experiments suggest that certain inhibitors may allow nitrate synthesis from humus nitrogen but not from ammonium that is two different populations are involved.

Nevertheless, direct and unequivocal evidence for the heterotrophic oxidation in soil is lacking, and pure culture investigations have revealed that no heterotrophic association can nitrify as quickly or as extensively as the two-genus autotrophic association. Nonetheless, the heterotrophs' inefficiency may be offset by their large numbers, and they may have some influence on the rate of nitrate synthesis.

Influence of environmental factors on nitrification

The rate of ammonium oxidation is influenced by physical and chemical factors. This is easily proved by the fact that nitrification rates in sterile soils receiving the identical inoculums vary by soil type. Because the responsible nitrifying species have a high degree of physiological similarity, environmental alteration has a significant impact on the end product's production. Little or no nitrate is discovered when the habitat becomes unfavourable, such as in acidic or anaerobic circumstances, whereas ammonium accumulates because ammonification is less sensitive to environmental change.

Acidity

A substantial link between nitrate production and pH has been discovered in several studies. Even in the presence of an adequate supply of substrate, nitrification occurs slowly in acidic environments, and the relevant species are rare or nonexistent at high acidities. Although nitrate can be found in fields down to pH 4.0 and sometimes lower, the rate drops dramatically below pH 6.0 and becomes negligible below 5.0.

Some soils nitrify at 4.5, while others do not; the difference could be due to acid-adapted bacteria or chemical variations between the two environments. Because acidity influences not only the transformation but also the microbial population, neutral to alkaline soils have the most. The pH levels required for these bacteria to flourish are influenced to some extent by the location from where they were isolated.

Liming acidic soils boosts nitrification, although it has little to no effect when the reaction is near the optimal pH. Indeed, in severely acidic, nitrate-free settings, liming may cause ammonium oxidation in areas where it has not previously happened. Because failure to nitrify is frequently caused solely by acidity, the problem can often be easily remedied by liming.

Oxygen

Aeration is an absolute necessity for all of the species involved. When microorganisms' oxygen supply is insufficient, little ammonium oxidation occurs, and the reaction eventually stops in the absence of oxygen. Artificial aeration increases bacterial activity and nitrate accumulation in controlled trials, whereas lowering the partial pressure of O2 below that found in air lowers nitrification.

Nitrification occurs even in submerged soils used for growing rice because O_2 diffusing from the atmosphere through the water phase keeps the upper few millimeters of soils oxygenated. Below the O_2-containing zone, no nitrate is synthesized.

Moisture

It has an impact on soil aeration. As a result, the microbial habitat's water state has a significant impact on nitrate production. Nitrification is reduced in high water logging conditions, which impede O_2 diffusion. In arid environments, on the other hand, bacterial proliferation is slowed by a lack of water. Irrigation boosts nitrifying population as well as nitrate biosynthesis. Varied soils have different ideal moisture levels, although nitrate emerges most frequently at one-half to two-thirds of the soil's moisture holding capacity.

Temperature

Temperature has a significant impact on nitrification, which occurs slowly below 5°C and quickly above 40°C. As the temperature rises from the lower extremes, the ammonium oxidation rate accelerates until the optimal range is reached. This range varies, owing to physiological changes in the prevalent bacterial strains, but the ideal temperature is usually between 30 and 35^0 degrees Celsius.

Crop

The amount and activities of the nitrifying microflora may be influenced by the type of crop grown. Although there are no preferential crop influences with many crops, a single plant species may have no effect on nitrification in one soil type while increasing nitrification in another. The relative abundance of ammonium over nitrate is a noticeable feature of many grasslands.

These characteristics have led to the hypothesis that the roots of grasses as well as those of other plants excrete compounds deleterious to nitrification. A striking effect of roots became apparent when trees and shrubs in a northern forest were cut and re-growth was prevented by herbicide application; significant quantities of nitrate accumulated in the soil and were transferred to underlying waters.

(a) This flush of nitrate may be a reflection of an inhibition of nitrification when the vegetation was present. No root exudates-no inhibitor of nitrification.

(b) Alternatively, it may merely signify that vegetation was able to remove the nitrate as readily as it was produced. No vegetative growth, no nitrate utilization, more nitrate accumulation leads to more leaching.

Seasonal effect

The seasonal influence is mostly determined by the interplay of moisture, temperature, cropping, and inorganic fertilisers. Furthermore, because plant uptake, microbial immobilisation, and leaching all reduce nitrate levels, the season of the year when nitrate is most abundant does not always correspond with the time of maximum microbial activity. In temperate zones, nitrate generation is fastest in the spring and autumn and slowest in the summer and winter, but annual moisture and temperature fluctuations can significantly alter the seasonal impacts.

Material undergoing decay

The rate of nitrate generation is also affected by the availability and type of substrates (materials). In near-neutral settings, nitrate is created more quickly from ammonium salts than from organic nitrogen molecules, whereas in acid soils, nitrate is formed faster from organic materials. The unanticipated discrepancy is due to an increase in alkalinity during ammonification, which makes the nitrifiers' environment more hostile.

i. When urea is digested by urease-producing microbes, the resulting ammonia raises the pH, hence urea causes more nitrate production in acid soils than $(NH_4)_2SO_4$.

ii. The application of fungicides reduces or eliminates the nitrifying species, allowing ammonium to remain and potentially become phytotoxic. The treated soil often reacts to inoculation after the pesticide has faded, albeit natural contamination would eventually overcome the deficiency.

iii. These organisms' nitrifying function is critical for plant nutrition availability. Nitrous acid (HNO_2) or nitric acid (HNO_3) is generated when ammonium salts are oxidised, and the pH drops. This increases the availability of particular ions to plants, which are generally difficult to obtain. Thus, nitrification has been linked to changes in soluble potassium, phosphate, magnesium, iron, manganese, and calcium concentrations.

Nitrate pollution

Nitrate, though an important ion in plant nutrition, may be a significant environmental pollutant. Recently a marked increase in addition of nitrogen to soil has been observed. This increase may attributed to:

- Use of large quantities of chemical fertilizers for crop production.
- The application to fields of vast amounts of animal manure from feedlots as part of disposal operations.
- Or the addition of digested sludge from cities and towns (to avoid polluting nearby water with the sludge) derived from sewage treatment plants.

Just to keep pace with a growing world population, increased input of fertilizer is applied and therefore there is enormous growth of fertilizer industry in the past few decades. Chemical fertilisers, which often comprise ammonium salts, anhydrous ammonia, or urea, are quickly nitrified, whereas organic nitrogen from carbonaceous wastes takes longer to convert to nitrate.Unfortunately, only the crop recovers a part of nitrogen thus added, and much of the unutilized nitrogen is oxidized to nitrate. This nitrate is then subjected to either leaching to the ground water or to denitrification.

In USA, many feedlots have tens of thousands of animals on small areas of land. The high cost of transporting manure to farmland distance away has made the issue of nitrate pollution serious from feedlots. Therefore, a feedlot of thousand animals may create significant pollution as the organic materials are mineralized and the resulting ammonium is oxidized to nitrate.

i. Nitrate travels through the soil profile until it reaches the groundwater.

ii. The subterranean water then is transported laterally to enter wells used for human drinking supplies.

iii. Or to surface waters that have the capacity to support algal blooms.

The movement of water to the ground water, wells or surface waters, is undesirable because of its potential role in:

a. Eutrophication

b. Methemoglobinemia in infants linked to drinking nitrate-rich water or eating nitrate-rich veggies

c. Animal methemoglobinemia and

d. Formation of nitrosamines.

Eutrophication

Lakes and rivers support certain population of algae or sometimes rooted plants, but the biomass is usually limited by the poor supply of inorganic nutrients. However, when additional nutrients enter, algae or rooted plants may flourish to create undesirable conditions. Their excessive growth depletes the oxygen dissolved in water. Thus, excessive loading (enrichment of water) with nutrients and algal growth causing reduction in O_2 level is called eutrophication. The nitrate entering from adjacent land areas after passing through the groundwater may promote modest to extensive blooms. Only a minimal concentration of nitrate is required, and lakes with 0.3 ppm nitrate-nitrogen may occasionally (but not always) support undesirable algal blooms if other nutrients are available.

- The use of inorganic or organic nitrogen in agriculture.
- Or the natural movements of nutrients from fields to inland waters.
- As well as the release of nitrogen when virgin soils are brought under cultivation, undoubtedly enhances the enrichment of these bodies of water in which this element scarcity limits the photosynthetic population.

Excessive growths of algae and rooted plants are undesirable for many reasons:

- Recreational uses of lakes are diminished
- The cost of water treatment rises
- Fish may succumb as the O2 is depleted when the algae die and are decomposed.
- The drinking water acquires offensive tastes and odours
- Navigation of small boats may even be hindered because of extensive networks of higher plants.

The disposal of solid or liquid sludge from municipalities or from large animals on land is advantageous because it provides the vegetation with essential nutrients and

Often leads to greater humus levels and Eliminates a bulky waste from the municipal treatment facility or the farm. This beneficial practice may also have disadvantages, one of which is the accumulation of nitrate to undesirable concentrations. Under such circumstances, reducing the rate of sludge application will decrease the amount of nitrate released from the microbial conversions so that groundwater pollution is minimized or overcome.

Methemoglobinemia

Methemoglobinemia disease arises when nitrate consumed with water, food or feed is reduced to nitrite in the gastrointestinal tract. The nitrite enters the blood stream and there reacts with hemoglobin to yield methemoglobin that leads to an impairment of O2 transport in the body. This process is not of consequence in adult humans but can be serious in infants, particularly those under 3 months old and in ruminants. This disease is also called blue baby syndrome.

Because the frequency of clinical cases is directly related with nitrate concentration in the water supply, the World Health organization and various countries have established standards for the quality of water designed for human consumption. The standard is usually 10 ppm nitrate-nitrogen, a level below which clinical methemoglobinemia in infants is extremely rare.

Animal methemoglobinemia

A few species of plants assimilate much nitrate from soil and store it in large quantities. These notorious accumulators are:

Vegetables: Beets, spinach, celery and lettuce.

Forage crops: Corn, sorghum, Sudan grass and oats.

In livestock herds, considerable deaths have occurred because the animals consumed feed obtained from plant species that stored the ion in appreciable amounts. Although, food-induced methemoglobinemia is almost unknown among babies. Active research is being conducted to find ways of reducing nitrate formation in soil under vegetables and to diminish accumulation in the crops that could present special problems.

Nitrosamines

Nitrosamines are formed in the soil by a simple condensation reaction involving a secondary amine (RNHR') and nitrite. The reaction can be carried out by microbial enzymes or non-enzymatically by organic materials present in soil. Interest in these chemicals arose dramatically when it became evident that many Nitrosamines are carcinogenic, mutagenic and teratogenic; that is they can induce cancer, mutations or abnormalities and sometimes death in foetuses.

Nitrite is not usually found in soil but sometimes it appears in large amounts and is constantly being generated. Secondary amines are not known to be common soil constituents but some pesticides are secondary amines. Plant residues having residual amount of pesticides often contain them and they can be produced microbiologically from other pesticides or natural products. This concern is supported by the findings that samples of soil in the laboratory do indeed form these toxicants and the plants can assimilate them which might be used by human as food. But to date no evidence exists that these chemicals are produced in soil under natural conditions.

Nitrification inhibitors

Nitrification may have considerable consequences. Although plants readily assimilate nitrate, the anion is also far more susceptible to leaching than ammonium so that the nutrient may pass through the soil out of the rooting zone.

- This leads to a loss of an essential element to be used for food production.
- It may also be responsible for water pollution because excessive nitrate in H2O can be hazardous to infants and livestocks.
- Excessive nitrate can enhance the growth of aquatic plants in nearby surface waters.
- The nitrate may also be converted to gaseous products due to denitrification that are unavailable for plant use.

For these reasons, a great deal of research has been devoted for finding inhibitory chemicals that might be mixed or applied together with ammonium or urea fertilizers to reduce the rate of nitrification. The compound of choice - must not be phytotoxic, nor can itself be an environmental pollutant. Many chemicals have been patented for this purpose, azide, chlorinated pyridines and pyrimidines, thiourea, cyanoguanidine, dicyandiamide (DCD), nitrobenzenes, substituted formamidines, N-alkyl-malemides, pyrazoles, thiadiazoles, triazoles, s-triazines and trichloroacetamides. The most promising from the practical standpoint are Dwell, 2-chloro-6 (trichloromethyl) pyridine (N-serve), ATC and DCD.

Remarkably low concentrations of these compounds inhibit the ammonium- and nitrite- oxidizing autotrophs but have no effect on heterotrophs that can participate in nitrification in culture. By suppressing nitrification, the amount of inorganic nitrogen lost through leaching and denitrification is diminished. Thus, more fertilizer nitrogen is available for plant assimilation and consequently crop yields can be increased with lower rates of fertilizer application.

Common name	Chemical	% inhibition (by 5 mg/kg soil) for 14 days
Dwell	Etridiazole	91%
N-serve	2-chloro-6 (trichloromethyl) pyridine	85%
ATC	4-amino-1,2,4-triazole	87%
DCD	Dicyandiamide	81%
AM	2-amino-4-chloro-6-methylpyrimidine	60%
ST	Sulpha-thiazole	52%
TU	Thiourea	2%

These nitrification inhibitors such as N-serve, ATC, DCD and Dwell are most widely used inhibitors to increase the efficiency of N-fertilizer by crop plants to maximize the crop yield.

Objective type questions

A. Fill up the blanks with suitable words

i. *Nitrosococcus*is ________________________shaped.

ii. *Nitrosomonas* and *Nitrobacter* are________________________in nature.

iii. The ratio of ammonium-nitrogen oxidised to CO_2-carbon for *Nitrosomonas* is __________________

iv. The ratio of ammonium-nitrogen oxidised to CO_2-carbon for *Nitrobacter* is __________________

v. The free energy efficiency of*Nitrosomonas* is typically__________________ percent

vi. *Nitrobacter's* free energy efficiency is estimated to be__________________ percent

vii. *Nitrobacter's*energy-producing process is suppressed by as little as __________________ ppm ammonium-nitrogen at pH 9.5.

viii. Nitrification is carried out in____________________ steps

ix. The presence of ____________________is mandatory for nitrification to happen.

x. The organisms that are abundantly involved in nitrification are __________________ and __________________

B. Multiple choice questions

i. *Nitrosomonas* and *Nitrobacter* are the organism responsible for

a. Nitrification b. Mineralization

c. Immobilization d. Denitrification

ii. 3 mole of ammonium ion on nitrification provides........mole of H^+ ions

a. Two b. Four

c. Six d. Eight

iii. The process in which Ammonium (NH_4^+) is converted to Nitrate (NO_3^-) is known as

a. Denitrification b. Immobilization

c. Nitrification d. Leaching

iv. Nitrate ion (NO_3^-) is susceptible to

a. Leaching
b. Denitrification
c. Nitrification
d. Both A and B

v. The optimum pH of nitrification is

a. 6.5
b. 7.5
c. 8.5
d. 9.5

vi. The optimum temperature for nitrification is

a. 0-10ºC
b. 10-20ºC
c. 15-25ºC
d. 25-35ºC

vii. Nitrification is performed by a small group of

a. Autotrophic Bacteria
b. Heterotrophic Bacteria
c. Fungi
d. Actinomycetes

viii. In dissimilatory denitrification, ______ serves as the electron acceptor in energy metabolism

a. Ammonium
b. Dinitrogen
c. Nitrate
d. Nitrite

ix. Which of the below is used to denitrify nitrates

a. Carbon
b. Aluminium
c. Iron
d. Copper

x. Nitrification is the process of turning__________into_____

a. NO_3^-. . . N_2
b. NO_2^-. . . NO_3^-
c. NO_3^-. . . NH_4^+
d. NH_4^+. . . N_2

Answer keys

A. Fill in the blanks

i. Spherical **ii.** Chemoautotrophic **iii.** 14-70:1 **iv.** 76-135:1 **v.** 5-14 **vi.** 5-10 **vii.**1.4 **viii.** Two ix.Oxygen **x.** *Nitrosomonas* and *Nitrobacter*

B. Multiple choice questions

i. **a** ii. **c** iii. **b** iv. **d** v. **c** vi. **d** vii. **a** viii. **c** ix. **a** x. **b**

C. Descriptive type questions

i. How nitrification may lead to environmental pollution?

ii. Briefly mention the role of nitification inhibitors in soil system?

iii. What do you mean by mineralization and immobilization in soils?

iv. Differentiate between autotrophic and heterotrophic nutrition?

v. Write the genera of organisms that participate in the nitrification process in soils.

11

Denitrification

Introduction

The various reactions of nitrogen (N_2) cycle, transform one form of the element to another. Mineralization liberates nitrogen in inorganic form, immobilization converts its back to an unavailable state, while nitrification changes the nitrogen from a reduced to oxidized state. Certain transformations of nitrogen lead to a net loss of element from the soil through volatilization. For the purpose of crop production, nitrogen volatilization has a deleterious influence because it depletes part of soils reserve of an essential nutrient. The sequence of steps that results in gaseous loss is called denitrification. So denitrification refers to the reduction of nitrates and nitrites to gaseous nitrogen products, principally di-nitrogen and nitrous oxide, coupled to energy production through oxidative phosphorylation. The ATP is generated during such electron transfer. It is an example of anaerobic respiration where electron acceptor other than O_2 is used. So if nitrification is considered reference point for initiation of N_2 cycle, denitrification is considered as termination.

Significant amount of inorganic nitrogen applied in fertilizers or formed from humus is recovered in the crop. Much of the element is removed by leaching and an appreciable amount disappears from the ecosystem entirely by volatilization. Three reactions of volatilization mechanism are possible:

(a) non-biological losses of ammonia, (in alkaline soil, > pH 9.0 or during rotting of manure)

(b) chemical decomposition of nitrite, and

(c) microbial denitrification leading to liberation of N_2, N_2O and sometimes NO.

Microbiology

Denitrifying bacteria comprise 0.1-5% of total bacterial population in soil. It represents a wide range of taxonomic groups. Soil microbial community contains several million denitrifiers per gram of the soil.

1. Population density of denitrifying bacteria can be estimated by using most probable number procedure. It is based on the conversion of nitrite or nitrate to gaseous end product in liquid culture.
2. Most accurate and sensitive method for enumerating bacteria is by determining encoding genome sequences and similarities in gene sequences of various denitrifiers present in soil sample.

Denitrifiers population data indicate high number of microorganisms present in the soil ecosystem that catalyze the process of denitrification. But in most soil, population densities of denitrifiers donot correlate with denitrification rate because (i) soil microbial community contains several million denitrifiers per gram of soil and (ii) these bacteria do not completely depend on the nitrate reduction for their growth.

General traits of denitrifiers

Denitrifiers are biochemically and taxonomically very diverse group of bacteria.

- Some denitrifiers are heterotrophic which derive energy from oxidation of fixed carbon sources.
- Some denitrifiers bacteria are chemocutotrophic when derived energy from hydrogen and reduced sulphur.
- Denitrifying bacteria are usually aerobic but when O2 is absent or anaerobic condition, then nitrate is used as electron acceptor for growth. Thus, active species grow aerobically in presence of O2 and anaerobically in absence of O2 but in the presence of nitrate. When nitrate acts as the electron acceptor during respiration instead of O2, it is called the nitrate respiration (anaerobic respiration).

During denitrification, the primary substrate is nitrate and end product is di-nitrogen. But some denitrifiers can reduce only nitrite i.e. nitrite dependent denitrification. Some microorganisms lack nitrous oxide reductase enzyme. So end product in such microorganisms is nitrous oxide (N2O). Thiobacillus denitrificans differ from other Thiobacillus bacteria in a special way that the can proliferate anaerobically in [presence of NO3.

Characteristics of efficient denitrifiers

- At least 80% of nitrate or nitrite should be reduced to di-nitrogen or nitrous oxide by denitrifying bacteria.

- There must be increased growth yield due to reduction of nitrate and nitrite to di-nitrogen.
- The conversion of nitrate or nitrite to nitrous oxide and N2 must occur at high rate. So the process should be a central reaction of cellular metabolism, not just side reaction.
- The presence of cytochrome c, d or disimillatory nitrite reductase should be demonstrated in microbial cell.

Genera of denitrifying bacteria

1. Archaea: *Halobacterium mansmorbii* was the first archaea identified to produce gaseous nitrogen oxide from NO_3^-. Hyperthermophile, strictly anaerobic *Ferroglobus placidus* respire NO_3 but not nitrite. But when supplied with NO_2, it produces substantial N_2O. Hyperthermophile is placed at the beginning of evolution of both archaea and bacteria. The occurrence of denitrification in hyperthermophiles can be viewed as indication of an early origin of denitrification process.

2. Organotrophs: Use organic compounds as electron donor while using NO3 as electron acceptor under anaerobic conditions. Although they respire aerobically in presence of O2.

Alcaligenes: commonly isolated from soil.

Bacillus: spore-former, fermentative, some species are thermophilic.

Rhizobium: symbiotic N_2-fixer with legumes. e.g. alfalfa and clover nodulating *R. meliloti* and *R. leguminosarum* bv. *trifolii*.

Azospirillum: associative N_2-fixer, fermentative. e.g. *A. lipoferum* and *A. brasiliense*.

Flavobacterium, Achromobacterium, Vibrio, Serratia.

3. Phototrophs: *Rhodobacter sphaeroides* sub. sp. *denitrificans*.

Rhodopseudomonas: anaerobic, reduce SO_4.

4. Lithotrophs or chemoautotrophs: use inorganic compounds as electron source for generation of energy.

Bradyrhizobium: use H2, also heterotrophic, symbiotic N2 fixer, isolated from soil.

Pseudomonas: use H_2, also heterotrophic. *Pseudomonas fluorescens, P. putida.*

Alcaligenes: use H_2, also heterotrophic.

Nitrosomonas: NH_3 oxidiser. *N. europaea*: produce N2O from NO2.

Paracoccus denitrificans: use H_2, facultative autotrophic and halophilic.

Micrococcus denitrificans: use H_2 and fixed CO_2 as energy source while O_2 and NO_3 as electron acceptor.

Thiobacillus denitrificans: chemoautotroph, use elemental S, sulfide and thiosulphate as energy source, also reduce nitrate to volatile product.

5. Obligate aerobic denitrifier: *Thiosphaera pantotropha:* S oxidiser, heterotrophic nitrifier, anaerobic denitrification.

Some heterotrophic nitrifying bacteria such as *Thiosphaera pantotropha* can nitrify and denitrify simultaneously. Such microorganisms may be valuable in bioremediation of ground waters and sewage effluents with high NH4+ and NO3 levels. In such cases, there should be no need to add energy yieklding substrate.

Pathways for biological reduction of soil nitrate

Soil fixed nitrogen resources may be conserved through both assimilatory and dissimilatory nitrate reductive process or they may be reduced by dissimilatory nitrate reduction. Assimilatory and dissimilatory nitrate reduction differ in ultimate fate of reduced N_2 atom.

(1) With assimilatory reduction, nitrate is reduced to ammonium, which is incorporated into cellular biomass. In contrast, for dissimilatory nitrate reduction, nitrogenous compounds accept electrons in support of cellular respiration. The final products di-nitrogen, nitrous oxide or ammonium are released from cell and accumulate in environment in concentration far beyond those necessary for biomass synthesis.

(2) Because assimilatory reduction of nitrate is linked to biomass production, it is catalyzed by all soil inhabitants capable of using nitrate as nitrogen source.

Three common microbial processes are classed under dissimilatory nitrate reduction. These processes can be distinguished by their respective products.

(i) Nitrite

(ii) Nitrous oxide and N2

(iii) Ammonium

The first dissimilatory reduction of nitrate involves simple reduction of nitrate to nitrite. Under anoxic (anaerobic) conditions, nitrate is reduced to nitrite by portions of general soil bacterial population. Then this nitrite diffuses from anoxic site where it is produced, to aerobic microsites, where nitrite oxidizers use it in energy generation.

Dissimilatory ammonium production and denitrification have several common characters. Both processes are coupled to organic matter decomposition, the nitrogen oxides that are reduced serve as terminal electron acceptors and growth yields of bacteria are increased as a result of energy provided by passage of electrons through an electron transport chain.

Enzymes involved in denitrification

The denitrification pathway involves four reductive steps and the corresponding enzymes are:

(1) ***Dissimilatory nitrate reductase (Nar):*** It is a membrane bound enzyme that contains Mo or Fe and labile S groups. It catalyzes the reduction of nitrate to nitrite with the generation of ATP. Both synthesis and activity of Nar is inhibited by O2. Recently, another nitrate reductase with low nitrate reduction activity has been found in the periplasm. This enzyme is termed as Nap.

(2) ***Nitrite reductase (Nir):*** The reduction of nitrite to nitric oxide is catalyzed by this enzyme. Two forms of this enzyme are known - one contains Cu (Cu-Nir) and other contains cytochrome c and d (heme-Nir, present in 2/3 microorganisms). Nitrite reductase is associated with periplasmic side of cell membrane. Nitrite reductase synthesis is repressed by O2. The presence of nitrate induces the expression of nitrite reductase.

(3) ***Nitric oxide reductase (Nor)*:** Nitric oxide is converted to nitrous oxide by activity of nitric oxide reductase (Nor). Nitric oxide reductase is membrane bound and contains cytochrome b and c. Because it is, a membrane bound enzyme, so electron transport associated with Nor enzyme is linked to ATP synthesis. The N=N bonds are formed in this step, however, the exact mechanism of this reaction is not known.

(4) ***Nitrous oxide reductase (NOS):*** This enzyme reduces nitrous oxide to di-nitrogen gas. Nitrous oxide reductase is a periplasmic protein that contains eight Cu atoms. Although nitrous oxide is not associated with cell membrane, energy production must be associated with the final reductive step. Denitrifiers use nitrous oxide as final electron acceptor. Synthesis of this enzyme is regulated by oxygen and nitrous oxide.

Biochemical properties of denitrification

During biological denitrification, nitrate is transformed to di-nitrogen by a series of reduction reaction as follows:

$$2HNO_3 \xrightarrow[-2H_2O]{+4H} 2HNO_2 \xrightarrow[-2H_2O]{+2H} 2[NO] \xrightarrow[-H_2O]{+2H} N_2O \xrightarrow[-H_2O]{+2H} N_2$$

Nitric oxide is written in written in brackets, because, it is not detected as free intermediate. There is also controversy whether NO is a true intermediate in the process or whether its presence represents a side reaction.

(A) ***Carbon and energy sources for denitrifiers:*** The electron for N_2O reduction may be provided by oxidation of a variety of fixed carbon compounds. Most commonly, investigators add glucose to soil samples as a carbon source in laboratory studies of denitrification processes. Many soil microbiologists do not consider denitrification to be a true anaerobic process, because the primary difference in nitrate respiration and O_2 based respiration is that in nitrate respiration electron transport process, nitrate is merely replacing O_2 as final electron acceptor. Denitrifiers are facultative anaerobes i.e. (i) they use molecular O_2 as terminal electron acceptor, (ii) if O_2 is not present then use NO_3^- as electron acceptor, (iii) these do not ferment organic substances in absence of molecular O_2. Denitrifiers also possess cytochromes that can transfer electrons to nitrogen oxides.

(B) ***Induction and synthesis of nitrogen oxide reductase:*** A modification of cellular metabolism of denitrifier growing in aerobic situation is that the enzymes required for denitrification are not synthesized in limited quantities. Nitrogen oxide reductase synthesis is repressed by molecular O_2. Once O_2 is depleted in microsite where denitrifiers reside, synthesis of these denitrifying enzymes is induced. So development of a totally functional denitrification system in aerobic ecosystem requires about 40 minutes to 3 hours.

Environmental factors controlling denitrification

1. ***Nature and amount of organic matter:*** Most members of soil denitrifying bacteria derive energy from oxidation of the fixed carbon sources. The stimulation of denitrification process by carbon amendment shows that it is a carbon-limited process. They are directly correlated. The increase in organic matter content also causes increase in the denitrification rate. Readily decomposable compounds such as simple sugars and organic compounds stimulate process more than less fermentable straw of grasses.

2. ***Nitrate concentration:*** Denitrification is an enzyme catalyzed reaction. So this reaction must follow Michalis-Menten kinetics. The addition of nitrate will increase denitrification rate until all enzyme will get saturated with nitrate. Then there is no further increase in denitrification rate.

3. ***Aeration or moisture:*** The biochemistry of denitrification shows that it is an anaerobic process. Therefore, oxygen represses the synthesis of reductases associated with denitrification. As in soil oxygen tension is reduced (because of slow rate of diffusion of O2 in H2O compared to air), then reduction of nitrate to nitrous oxide and di-nitrogen increases. Thus effect of aeration on denitrification rate in soil is highly related to soil moisture.

In general, denitrification is not detected below 60% water holding capacity (WHC). The value of denitrification rate generally correlates with soil moisture. The increase in denitrification rate is due to increase in number of anaerobic microsites.

4. ***Soil reaction (pH):*** Although denitrification occurs within pH range of 3.9-9.0. Maximum nitrogen oxide reduction occurs from 7-8 pH. The rate of denitrification decreases with decrease in pH of the soil.

- N_2 tends to be the dominate product above pH 6.0 to 7.0.
- Under low pH (6.0-6.5), more nitrous oxide (N_2O) is produced and frequently makes more than half of the nitrogenous gases evolved from acid habitat.
- Below pH 5.5, the NO is produced.

Bacterial genera, Pseudomonas, Micrococcus, Achromobacter and Bacillus are not present below pH 5.5.

5. ***Temperature*****:** Denitrification is markedly affected by temperature. The transformation proceeds slowly at 2^{0}C, but increase in temperature will enhance the rate of denitrification. The optimal temperature for denitrification reaction is 25^{0}C and above. The transformation is still rapid as elevation of temperature occurs and will proceed till 60-65^{0}C. But it doesn't occur at 70^{0}C, because of its effect on solubility of O_2 and N_2O, effect on diffusion coefficient and effect on O_2 consumption activity of heterotrophs. The rapid release of N is more at elevated temperature, suggesting the involvement of active thermophillic flora.

6. Cropping: Cropping serves to reduce volatilization losses through denitrification. This retardation in cropped land suggests that vegetation is competing with the microflora for the supply of inorganic nitrogen, thereby reducing the amount of nitrate acted on microbiologically.

By contrast, when the nitrate supply is adequate, plants promote the microbial release of volatile nitrogenous products and also the relative amount of N_2 and N_2O that are liberated. This enhancement may be result of the roots providing exudates that support proliferation of the denitrifying bacteria and/or the diminished O_2 tension in the root region or both.

Quantification of nitrogen losses from an ecosystem via denitrification

Quantification of N losses from an ecosystem is not easy, because of the following reasons:

1. N_2 is omnipresent in nature, so its release in field is not simple to demonstrate quantitatively.
2. Volatilization of N_2 is very slow process. So precise measurement of gas evolution is difficult.
3. Soil often contains excess of 2000 lb/acre of N_2. So measurement of losses 20-40 lb/acre is very difficult.

Nitrogen losses in field can be determined by changes in N content of soil with time. The various techniques used are:

1. Nitrogen balance studies
2. Use of N isotopes
3. Disappearance of N oxide
4. Acetylene block method

1. Nitrogen balance studies

In this process, N_2 is considered as uncountable N.

Denitrifying N_2 = (Pre-existing N in soil + N added as fertilizer + bio-fixed N) - Total fixed N in the soil.

It is based on three assumptions:

a. The entire organic and fixed N should be accurately quantified.

b. The assay procedure applicable for quantifying soil N compounds is sufficiently sensitive.

c. The sole means of gaseous N loss from soil is denitrification.

Limitations

i. Inability to accurate account for all fixed N_2 inputs and losses from soil causes variability in N_2 balance data.

ii. Applications of assay procedure to soil nitrogen reserves are limited by fact that changes in N reserves are commonly less than the precision of assay method.

iii. There are at least 3 major routes of N_2 volatilization:

- Non-biological loss of NH_3
- Chemical decomposition of nitrite
- Denitrification as di-nitrogen and N_2O (nitrous oxide)

2. Use of isotopes

Short-lived radioactive N isotopes have been used to quantify denitrification process. Due to short half-life of radioactive isotopes, the heavy isotope ^{15}N is more amenable to long-term evaluation of denitrification in soil. For these studies, ^{15}N labelled fixed N sources are amended to soil and then this ^{15}N labelled di-nitrogen and nitrous oxide is quantified.

Assumption

The soil amended with ^{15}N labelled N is homogeneously mixed with native soil N resources and thus of equivalent availability to denitrifier population.

Limitation

Because of heterogeneity of soil system, uniform mixing of an external nitrogen supply with indigenous N does not occur and thus two N sources are not equivalent available to denitrifier population.

3. Disappearance of N oxides

Since nitrate, nitrite and nitrous oxides are readily extracted from soil and quantified. So the rate of change in the concentration of these chemicals can serve as indication of denitrification rates.

Assumption

It is based on assumption that sole product of nitrate and nitrite reduction are gaseous.

Limitation

The change in concentration of these chemicals is very little (µg/gram of soil). So exogenous supply of nitrous oxide must be added in order to have meaningful estimate.

4. Acetylene block method

- Place 10 g of soil into 60 ml serum bottle.
- Add H_2O or KNO_3 solution to soil.
- Purge and evacuate bottle with argon or argon + 10% acetylene.
- Assay headspace N_2O with gas chromatography procedure.

Assumption

Sufficient acetylene is present in all microsites of system to inhibit the totality of nitrous oxide reductase present.

Limitation

i. It commonly involves its application under field conditions.

ii. Complication associated with long-term use of acetylene inhibition procedure is that, it is toxic to ammonium oxidizers, thereby inhibit natural source of nitrate for denitrification process.

Environmental implications of N_2O formation

At one time, the accumulation of nitrous oxide in reaction vessel where biological denitrification was taking place was ignored, but now it is known that N_2O is common product of denitrification. For example, fluxes of N_2O from soil have been shown to range from 7 to 165 kg N_2/ha/year in drained histosoles in South Florida. Annual losses from irrigated soils in California range from 19.6 to 41.8 kg N_2/ha/year. Lesser yield has been detected from cropped soils in Colorado as well as agricultural soils in USA. Uncropped native ecosystems also produce more quantities of nitrous oxide during plant growth seasons. These data show that nitrous oxide productions by soil processes are highly variable. Soil properties controlling the quantity of this product by soil microbes include:

i. Soil redox potential

ii. Moisture tension

iii. Nitrate concentration

iv. O_2 concentration

v. Time of day when measurements are taken. Generally, the proportion of gaseous nitrogen products composed of nitrous oxide increases with increasing soil acidity, reduced soil temperature and augmented soil nitrate level.

Environmental pollution caused by denitrification

(i) H_2O pollution: The N_2O produced is not fully reduced to N_2. Some part of it leaches down and this reaches to ground water level and cause water pollution which is cause of:

(a) Eutrophication

(b) Infant methemoglobinemia: After drinking NO_3 polluted water, NO_3 gets converted to NO_2 in gastrointestinal tract which combines with hemoglobin to make methemoglobin. It causes block in O_2 transport in body.

(c) Animal methemoglobinemia: NO_3 accumulates in forage crops and high NO_3 content are toxic to cattle and sheep. It is associated with nitrate rich water or vegetables.

(ii) Green house effect leading to global warming

Like CO_2, CH_4 and other gases, N_2O is also good absorber of infrared radioations. Therefore, it leads to global warming due to green house effect.

(iii) Ozone depletion

Nitric oxide and nitrous oxide (N_2O) produced during denitrification is the cause of O_3 depletion. N_2O is O_3 depleting gas.

Formation of ozone

$$O_2 \xrightarrow{light} 2O^* + 2O_2 \longrightarrow 2O_3$$

(reactive oxygen)

Reaction involved during O_3 depletion:

$$N_2O + O^* \longrightarrow 2NO$$

$2NO + O_3 \longrightarrow NO_2 + O_2$

$NO_2 + O^* \longrightarrow NO + O_2$

Therefore, formation of N oxides and depletion of ozone layer occurs. Not all the nitrous oxides evolving from soil surface results from activity of denitrifiers, a variety of soil heterotrophs as well as nitrifiers, also produce this gas as a metabolic by product.

Objective type questions

A. Fill up the blanks with suitable words

i Denitrification refers to the reduction of.........................and .... to gaseous nitrogen products

ii Denitrifying bacteria comprise .... of total bacterial population in soil.

iii Population density of denitrifying bacteria can be estimated by using ....procedure.

iv When nitrate acts as the electron acceptor during respiration instead of O2, it is called the ....

v Maximum nitrogen oxide reduction occurs from .... pH

vi Denitrification is not detected below ....water holding capacity

vii Drinking water with high nitrate can cause a potentially fatal disorder , particularly to the infants,called............................

viii Nitric oxide is converted to nitrous oxide by activity of the enzyme ....

ix Heterotrophic nitrifying bacteria such as .... can nitrify and denitrify simultaneously.

x ....is O_3 depleting gas

B. Multiple choice questions

i. Denitrifying bacteria are

a. heterotrophic b. chemoautotrophic

c. None of this d. Both of a and b

ii. During denitrification, the primary substrate is

a. Nitrogen b. nitrate

c. ammonia d. None of this

iii. *Halobacterium mansmorbii*which produce gaseous nitrogen oxide from $NO_3^-$was

a. Bacteria b. Fungus

c. Archaea d. None of this

iv. Phototrophs that are able to denitrify

a. *Flavobacterium* b. *hodobacter*

c. *Rhodopseudomonas* d. Both b and c

v. Nitrate is reduced to ammonium through the process

a) Mineralization b) Denitrification

c) Assimilatory reduction d) None of this

vi. Dissimilatory nitrate reductase is a membrane bound enzyme that contains

a) Mo and Ni

b) Zn and Cr and labile S groups.

c) Mo or Fe and labile S groups.

d) None of this

vii. In presence of O_2Nitrite reductase synthesis

a. Supressed b. Enhanced

c. Have no effect d. Not known

viii. Nitrous oxide reductase is a periplasmic protein that contains eight atoms of

a. Zn b. Fe

c. Mn d. Cu

ix. Denitrifiers are

a. Obligate anaerobes b. Obligateaerobes

c. Facultative anaerobes d. Both b and c

x. Under low pH (6-6.5) denitrification results production of more amount of

a. N_2O b. NO

c. NO_2 d. NH_4

Answer keys

A. Fill in the blanks

i. nitrates and nitrites **ii.** 0.1-5%, **iii.** most probable number **iv.** nitrate respiration or anaerobic respiration. **v.** 7-8 vi. 60% **vii.** Methemoglobinemia **viii.** nitric oxide reductase, **ix.** *Thiosphaera pantotropha* **x.** N_2O

B. Multiple choice questions

i. **d** ii. **b** iii. **c** iv. **d** v. **c** vi. **c** vii. **a** viii. **d** ix. **c** x. **a**

C. Descriptive type questions

i. What is denitrification? Named the enzymes involve in the denitrification process.

ii. How environmental factors control denitrification?

iii. Write down the characteristics of efficient denitrifiers

iv. Elaborate various methods for Quantification of nitrogen losses from an ecosystem via denitrification?

v. What are the effects of denitrification on environment?

12

Rhizobium-legume Symbiosis Formation of Nitrogen-fixing Nodule

The excessive and imbalanced use of nitrogenous fertilizers in the last few decades to meet the challenges of growing food demand of Worlds population has led to serious consequences of soil and water pollution, leaching of toxic nitrate into ground water and volatilization of nitrogen oxides into the environment. Therefore, judicious management of nitrogen in the environment is very much essential in the development of more productive and sustainable agriculture. Rapid declining of non-renewable energy sources viz. petroleum products, natural gas as well as energy crisis in the recent days, release of several pollutants during production of fertilizer N has also necessitated the development of alternative and/or supplemental technologies to obtain the fertilizer N for crop production.

Reduction of atmospheric elemental N to ammonia through biological means can only be performed by some prokaryotes which provide N input into the soil. Biological N_2 fixation can occur through symbiotic relationship or in association with crop plants as well as in free-living state. However, the most important N_2-fixing systems are the symbiotic associations from an ecological and agricultural point of view. Few important N_2-fixing symbiotic associations are *Rhizobium*-leguminous plants, *Frankia*-actinorhizal plants, *Anabaena-Azolla* symbiosis and lichen symbiosis involving cyanobacteria which have been studied in great extent. In the symbiotic association the host plant supplies the bacteria with carbon substrates and necessary nutrients essential to support bacterial growth and to provide energy as well as protection from oxygen for N fixation to happen. The legume-*Rhizobia* symbiosis can contribute over 100 million metric tons of fixed nitrogen per year to the global nitrogen economy which accounts for more than 65% of the nitrogen used in agriculture.

Generally, host, microsymbiont and environmental factors influenced the rates of N_2 fixation. On an average different symbiotic association viz. legume-*Rhizobium, Frankia*-actinorhizal and *Anabaena-Azolla* contributes 24-584, 2-362 and 45-450 kg N ha^{-1} annually, respectively. Such symbiotic associations render the plants to grow in N deficient soils as well as reduces its losses through various means viz. denitrification, volatilization and leaching, thus helps in achieving sustainable agricultural system. Therefore, growing legumes conserves soil N and augments soil N pool for the succeeding crops (non-leguminous) in rotation. The process of symbiosis is thus important in providing high protein diet and also serves as cover crops, green manure and forage crops for livestock.

Host specificity of Rhizobium species and bacterial promiscuity

The host plants nodulated by *Rhizobium* species belong to the family Leguminosae (Fabaceae), only with the exception of the genus *Parasponia* belonging to the family Ulmaceae. The family Leguminosae comprised of three sub families namely Caesalpinoideae, Mimosoideae and Papilionoideae, these three subfamilies contains genera that are nodulated by rhizobia. The quantity of nodulated species (%) under each of the three subfamilies also varied to a great extent. Out of the 748 genera (approx.) and about 19000 legume species, nearly 49% (genera) and 16% (species) have so far been examined for nodulation and nearly 41% (genera) and 14% (species) are reported to bear nodules. The most primitive subfamily Caesalpinoideae includes many non-nodulating genera whereas the most advanced subfamilies Papilionoideae and Mimosoideae are having very rare non-nodulating species in it, indicate the development of symbiosis was started at a relatively late stage during legume nodulation. The family Ulmaceae contains a species *Parasponia* which can form N_2-fixing nodules with a variety of rhizobial strains, which can also nodulate, some other legumes. In addition, few aquatic and water tolerant legumes also bear N_2-fixing nodules on root as well as on stem. This peculiarity is restricted to species of *Sesbania, Aeshynomene* and *Neptunia, Arachis hypogaea* and *Vicia faba* cv. Throws Ms.

Eleven genera of root-nodule bacteria have been identified so far according to present taxonomic classification of which *Rhizobium* includes the fast-growing species whereas, *Bradyrhizobium* includes slow-growing species, on the other hand, *Azorhizobium* includes the fast-growing species capable of forming both root and stem nodules on *Sesbania,* a tropical waterlogged legume (Table 1).

According to Chen *et al.* (1995) a separate genus named *Mesorhizobium* has been defined between the slow-growing *Bradyrhizobium* strains and the typical fast-growing *Rhizobium* strains to indicate a growth rate intermediate and subsequently, it was denoted as a phylogenetic position for rhizobia.

Table on Current status of rhizobial taxonomy and nodulation host range among rhizobium strains.

Genus	**Species**	**Host**
Allorhizobium	*A. undicola*	Neptunianatas
Azorhizobium	*Azorhizobium caulinodans*	Root and stem nodules on Sesbania
Bradyrhizobium	*B. japonicum*	Soybean (Glycine max)
	B. elkanii	Soybean (Glycine max)
Mesorhizobium	*M. ciceri*	Chick pea (Cicer arietinum)
	M. mediterraneum	Chick pea (Cicer arietinum)
	M. loti	Trefoil (Lotus sp.)
	M. chacoense	Propopis alba
	M. plurifarium	Acacia, Leucaena (Leucaena leucocephala)
Methylobacterium	*M. nodulans*	Crotolaria pedocarpa
Rhizobium	*R. elti, R. gallicum*	Bean (P. vulgaris)
	R. leguminosarum	Pea (Pisum sativum), Vetch (Vicia sativa)
	R. trifolii	Clovers (Trifolium sp.)
	R. tropici	Bean (P. vulgaris), Leucaena
	R. hainanense	Desmodium, Stylosanthes, Tephrosia
Sinorhizobium	*S. meliloti*	Alfalfa (Medicago sativa), Sweet clover (Melilotus alba) Fenugreek (Trogonellafoenumgraceum)
	S. fredii	Soybean (Glycine max)
	S. terangae	Acacia, Sesbania
	S. arboris	Acacia senegal, prosopis chilensis
Burkholderia	*B.* sp.	Aspalathus carnosa
Ralstonia	*R. taiwanensis*	Mimosa

The legume-rhizobia symbiotic relationship is quite complex and the degree of host specificity varies greatly among the rhizobia. For example, some strains like *R. meliloti* and *R. Leguminosarum* bv. *Trifolii* are known to have a very narrow host range. Some rhizobia from diverse taxonomical origin includes *Sinorhizobium saheli, Sinorhizobium teranga* bv. *sesbaniae* and *Azorhizobium caulinodans* also shows a specific host range limited mainly to *Sesbania* spp. However, *Rhizobium* sp. strain NGR 234 having a very wide host range which can nodulate more than 136 different legume genera, varying from woody plants to annuals and to a non-legume *Parasponia*.

Strains of both *Rhizobium* and *Bradyrhizobium* are reported to infect some legume hosts and form nodules with them viz. *B. japonicum* and *R. fredii*, both of them can nodulate soybean crop. Cowpea, green gram, black gram and pigeon pea etc. hosts of cowpea miscellany group are nodulated by both fast-growing as well as slow-growing rhizobia. Both fast- and slow-growing *Lotus* rhizobia (R. loti) were reported to nodulate *Lotus pedunculatus* (Big trefoil) and *L. corniculatus* (Birds foot trefoil) legume species grown in low fertility pastures. Some unusual specificity for legumes was also found with an Asian species *Sinorhizobium fredii*. This microsymbiont has been found to nodulate over 50 diverse legumes, in addition to its original host soybean (*Glycine max*) making it as symbiont with a considerably wide host range, strains *S. fredii* viz. USDA257, on the other hand, are highly specific for some cultivars of soybean.

Almost all the characterized rhizobial strains was derived from the narrow range of cultivated species of leguminous plants and all such rhizobial isolates belong to three distinct branches within the alpha-2 subgroup of *Proteobacteria*. Each of the rhizobia are phylogenetically intertwined with non-symbiotic bacteria. The genus *Rhizobium* has the largest branch that nodulates peas and clovers, whereas alfalfa (lucerne) nodulated by *Sinorhizobium*which isclosely related to *Agrobacterium* and to *Brucella*. The second group of rhizobial strains includes the genus *Bradyrhizobium*that nodulate soybeans, lupin and many tropical legumes and is closely related to *Rhodopseudomonas*. *Azorhizobium* is the third rhizobial group which is closely linked to the chemoautotroph *Xanthobacter*. Recently two more (completely new) rhizobial groups have been reported, which are i) *Methylobacterium nodulans* isolated from *Crotolaria* nodules represents a fourth class of alpha-2 subgroup of *Proteobacteria* and ii) Burkholderia isolated from the nodules of *Aspalathus*, a member of the distant beta-subclass of *Proteobacteria*.

Nodule formation on legume roots

The N_2-fixing mutualistic associations between plants of the *leguminosae* familyand the soil bacteria belonging to the rhizobacteria groups viz. *Azorhizobium, Bradyrhizobium, Mesorhizobium* and *Rhizobium* (collectively called rhizobia) contribute substantially to crop productivity. The symbiotic association between legume plants and rhizobacteria also offers an interesting model to study and investigate the various complex mechanisms that control plant cell division and nodule development.

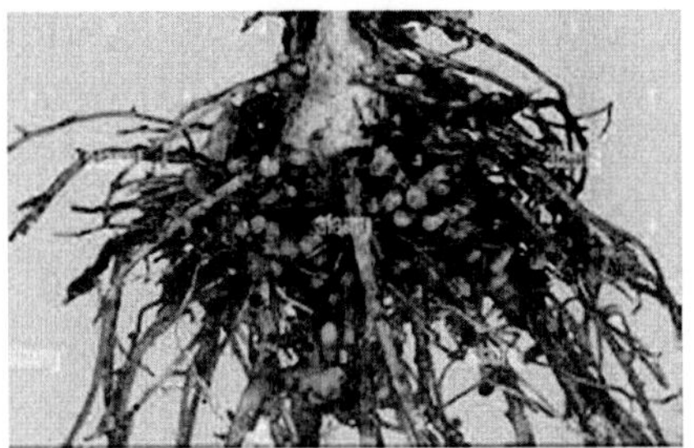

Fig. 1: Nodulated root of chick pea (*Cicer arietinum*)

The sequence of development and morphology of root nodules exhibit considerable diversity in different legume species, reflecting wide variation of the legume species in their characteristic symbiotic interaction. However, the overall mechanism for nodule induction seems to be identical for all forms of rhizobia. The four anatomical attributes which are characteristic of the nodules of different legume species are i) induction of plant meristem involving cortical cells of roots, ii) bacterial (Rhizobium) invasion to root tissue and cell, iii) central tissue development having reduced supply of atmospheric oxygen where N_2 fixation takes place and iv) peripheral vascular tissue development outside of the infected tissue but within the nodule endodermis.

The detailed process of nodule formation and function and the nodulation process can be arbitrarily divided into four stages:

- Nodule initiation
- Nodule invasion
- Nodule maturation and differentiation
- Nodule senescence

Nodule initiation

The free-living rhizobia, in absence of the host are in their saprophytic phase and do competition with other soil microbes for limited sources of soil nutrients. The rhizobia population densities in soil are generally highwhen legumes consist of considerable part of the plant community, suggesting the symbiotic association is critical for the establishment of a substantial saprophytic population of rhizobia in the soil. Naturalized rhizobia populations as well as inoculant rhizobia populations are known to differ in their tolerance to the major environmental factors, thus affecting their survival and persistence in the soil system.

Rhizobia respond to positive chemotaxis to root exudates containing flavonoids, amino acids, organic acids and sugars and move towards localized sites on legume roots. Generally, rhizobia attach to root hairs all throughout the root, however, the root hairs just behind the apical meristem at the site of emergence are most responsive to *Rhizobium* infection.

The rhizobia in the infected root zone attached to the surface of root hair either through an acidic extracellular polysaccharide or via specific calcium-dependent protein, rhicadhesin, cellulose fibrils and legume root lectin.

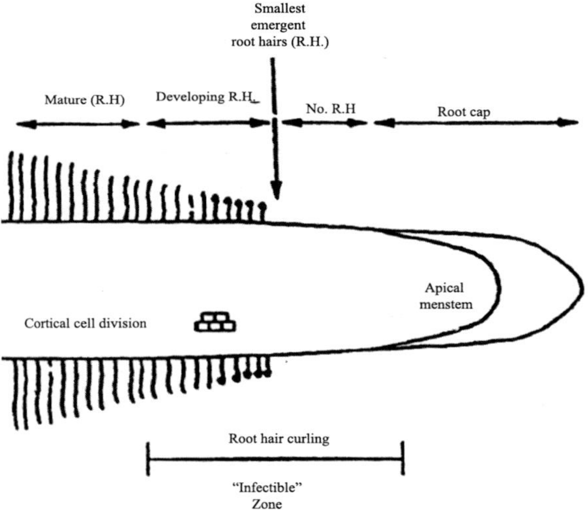

Fig. 2: Seedling root showing infectible zone.

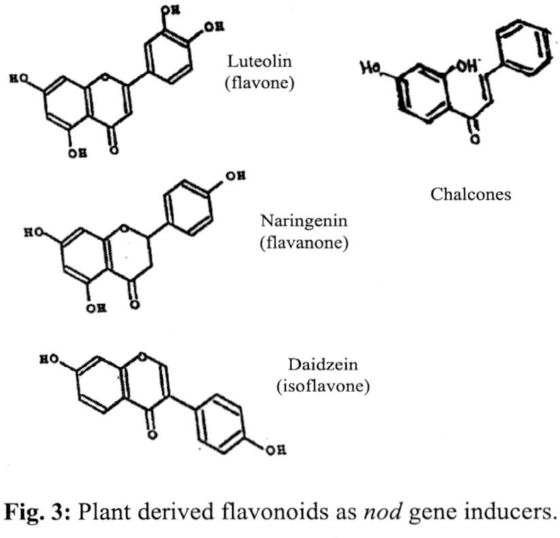

Luteolin
(flavone)

Chalcones

Naringenin
(flavanone)

Daidzein
(isoflavone)

Fig. 3: Plant derived flavonoids as *nod* gene inducers.

Nodule initiation step is characterized as a two-way molecular conversation between host plant and the *Rhizobium*.

i. The signal compounds are released by host legume. The specificity in induction of rhizobial nodulation genes by specific flavonoid signal molecule forms the first host specific step in nodule initiation. The transcription of *Rhizobium/Bradyrhizobium* nodulation genes is induced by the flavonoids in root exudates, which are not generally expressed or expressed at very low levels in free-living rhizobia in the absence of legume. Flavones and flavonones are usually the inducers of fast-growing rhizobia whereas isoflavones are often inducers of *Bradyrhizobium spp.*

Flavone: Luteolin is inducer for R. meliloti

Flavonone: Naringenin and hesperitin are inducers for R. leguminosarumbv. viciae.

Isoflavone: Diadzein and genistein are inducers for B. japonicum

ii. The expression of nodulation (nod) genes is activated by these signal compounds. The positive regulatory protein NodD is bind by the flavonoids and isoflavonoids secreted by legume roots which result in the transcriptional activation of other nod genes.

Nodulation (Nod) factors are synthesized and signal exchange occurs from Rhizobia to plant. Three types of nod genes viz. common nod genes, host-specific nod genes and regulatory nod genes of Rhizobia are required for infection and nodulation. Their products are required for the biosynthesis and secretion of nodulation (Nod) factors or lipochitin-oligosaccharides (LCO) that trigger nodule organogenesis in the plant host. In the production of basic lipochitin-oligosaccharide molecule the common nod genes are involved, whereas various substituents at the reducing or non-reducing ends of nod factors are added by the host specific nodulation genes and its mutations may cause either a delay in nodulation or a change in host range.

Nodulation factors cause morphological changes in plant roots leading to nodule organogenesis.

The infecting rhizobia produces lipo-oligosaccharides (Nod factors) in compatible host which cause characteristic curling and deformations of root hair and cortical cell divisions. Such deformations of root hair in various legumes may develop various structures viz. corkscrews, branches, twists, spirals and shepherd's crook.

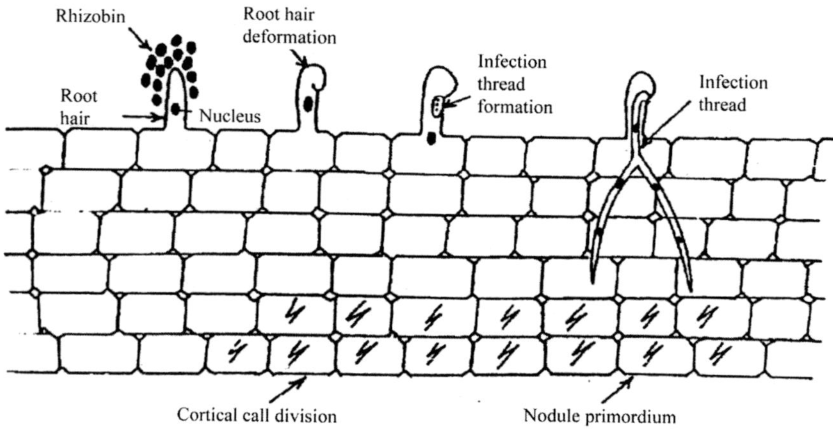

Fig. 4: Nodule initiation and nodule invasion stages during nodule organosgenesis.

Nodule invasion

The root hair cells in legumes are invaded by the rhizobia by producing a host-derived infection thread initiated from the most prominently curled regionas an invagination of the root hair cell membrane. Then the rhizobia travel down to the root hair to the underlying cortical cell layers through the inwardly growing tubular tunnel-like infection thread. The infection thread containing the rhizobia are embedded in a mucigel composed of different compounds viz. cell wall polysaccharides, plant-derived matrix glycoprotein and rhizobial exopolysaccharides. The infection thread then moves towards the newly developed nodule primordium formed by mitotic activity in the root cortex due to rhizobial nod factors and subsequently develops into the nodule meristem.

Through the continued growth of infection threads towards the nodule primordia, bacterial invasion to the nodule spreads from cell to cell. Infection threads ramify and penetrate individual nodule cells in the root cortex to form a new structure, the infection droplet. Through the process of endocytosis, rhizobia from each infection droplets are released into nodule tissue cells to occupy an organelle-like cytoplasmic compartment, named symbiosome, bounded by a plant-derived peribacteroid membrane.

Nodule maturation and differentiation

The bacteria enclosed in the peribacteroid membrane continue to divide into several thousand rhizobial cells in the cytoplasm of each infected plant cell. In this way the infected cells are completely filled with bacteria which have differentiated into their pleiomorphic endosymbiotic bacteroids in the late symbiotic zone. Such bacteroidsspecifically express genes responsible for N_2 fixation by reducing atmospheric nitrogen with the help of the nitrogenase enzyme. The N_2 thus fixed from the atmosphere is used by the plant as nitrogen source and plant in turn provide photosynthates and amino acids as C source, energy and nitrogen sources.

Nodule senescence

The programmed senescence of nitrogen-fixing bacteroidsis an integral part of the development sequence in intermediate nodules.Growth and division of bacteroids is stopped at this stage and lysis of N_2-fixing bacteroids and the host cells also occurs.After demise of the nodule, the bacteria enclosed in it are positioned in a way so that it can utilize the nutrients of the senescing nodule tissues for rapid proliferation.Nevertheless, the population of nodule-derived rhizobia entering the soil becomes low as most of the cells are destroyed along with the plant cells during nodule senescence and also, the differentiated bacteroids could not be able to switch from biotrophic to saprotrophic life in the soil.

Strategies for improving N_2 fixation in Rhizobium-legume symbiosis

Significant efforts have been made recently to improve the efficiency of biological nitrogen fixation as symbiotic as well as free-living microorganisms have the tremendous potential to substitute a considerable partof the mineral nitrogenous fertilizers produced industrially. However, most of the efforts for improving the N_2 fixation have been made on symbiotic N_2-fixing bacteria of the genera *Rhizobium* and*Bradyrhizobium*as they form symbiotic associations with legume crops having agronomicimportance.Alternative approach includes the genetic manipulation of non-legume crops to incorporate *nif* genes from bacteria and/or extension of the host range for symbiotic associations between rhizobia and non-legumes.

(i) Enhancement of nodulation and broadening of host range

   a. Transfer of symbiotic plasmid

   b. Transfer of cloned nod genes

(ii) Enhancement of nitrogen fixation

(iii) nodulation and nitrogen fixation in non-legume hosts.

Objective type questions

A. Fill up the blanks with suitable words

i. In the symbiotic association the host plant supplies the bacteria with ....and.......................................essential to support bacterial growth

ii. Rhizobium species belong to the family ....

iii. For *Bradyrhizobium* species the suitable host plant is ....

iv. Pea (*Pisum sativum)* is a suitable host plant for the species of Rhizobium ....

v. Alfalfa (lucerne)is nodulated by ....

vi. Specific calcium-dependent protein,...............................is involved in attachment of rhizobia to the roots

vii. The transcription of Rhizobium/Bradyrhizobium nodulation genes is induced by the ....in root exudates,

viii. The bacteria enclosed in the ....membrane continue to divide into several thousand rhizobial cells in the cytoplasm of each infected plant cell.

ix. The infection thread containing the rhizobia are embedded in a ....composed of different compounds

x. Healthy root nodules are pink in colour due to the presence of ....

B. Multiple choice questions

i. Which of the following is required for the action of the nitrogenase enzyme?

a. Light
b. High input of energy
c. Super oxygen radicals
d. Both of a and b

ii. What is the function of leghaemoglobin?

a. Oxygen removal
b. Inhibition of nitrogenase activity
c. Expression of nif gene
d. None of this

iii. Which of the following is correct for nitrogen fixing bacteria?

a. They convert free nitrogen to nitrogen compounds

b. They oxidize ammonia to nitrates

c. They reduce nitrates to free nitrogen

d. None of this

iv. This element plays a key role in the nitrogen fixation.

a. Zn   b. Cu

c. Mo   d. Both b and c

v. Cells where nitrogen fixation takes place in Nostoc are known as

a) Heterocysts   b) Akinetes

c) Nodules   d) Hormogonia

vi. What is the first stable product of nitrogen fixation in the root nodules of leguminous plants?

a) NO_3^- b) NO_2^-

c) Ammonia   d) None of this

vii. An aquatic fern that involves in nitrogen fixation in symbiosis with *Anabaena is*

a. Azolla   b. Salvia

c. Salvinia   d. All of this

viii. Industrial nitrogen fixation is carried out by

a. Friedel Crafts reaction   b. Helmonts process

c. Haber process   d. None of the above

ix. All are free living N2 fixers except

a. Rhizobium   b. Azotobacter

c. Rhodospirillum   d. Clostridium

x. To fix one molecule of Nitrogen number of ATP molecules are required as

a. 6   b. 12

c. 16   d. 20

Answer keys

A. Fill in the blanks

i. Carbon substrates and nutrients **ii.** Leguminosae **iii.** Soybean (Glycine max) **iv.** Leguminosarum **v.** *Sinorhizobium* sp. **vi.** rhicadhesin **vii.** flavonoids **viii.** peribacteroid, **ix.** mucigel **x.** leghaemoglobin

B. Multiple choice questions

i. **b** ii. **a** iii. **a** iv. **c** v. **a** vi. **c** vii. **a** viii. **c** ix. **a** x. **c**

C. Descriptive type questions

i. What is Nitrogen fixation? Named the different nitrogen fixation process?

ii. Write the difference between symbiotic and free living N2 fixers?

iii. Write down important rhizobium genus with nodulation host range?

iv. Write the difference between determinate and indeterminate root nodules?

v. Elaborate rood nodulation process in brief

13

Microbial Transformation of Phosphorus

Introduction

Phosphorus is only second to nitrogen as a mineral nutrient required for plants, animals and microorganisms. It is a major constituent of nucleic acids in all living systems essential in the accumulation and release of energy during cellular metabolism. Phosphorus is found in soil, plants and microorganisms in number of organic and inorganic compounds. It is the second major inorganic nutrient after nitrogen in term of quantitative requirement for both crop plants and microorganisms. This element may be added to soil in the form of chemical fertilizers, or it may be incorporated as leaf litter, plant residues or animal remains. In cultivated soils it is present in abundance (i.e. 1100 kg/ha), but most of which is not available to plants, only 15% of total soil phosphorus is in available form. Thus, phosphorus occupies a critical position both in plant growth and in the biology of soil.

The process associated with transformation of phosphorus occurring in soil is governed by the principles previously described for other soil-based biogeochemical cycles. For example, (i) mineralization of organic compounds with the release of inorganic phosphate and (ii) assimilation or immobilization processes converting the inorganic available anion into cell components (microbial biomass and plant) are catalyzed by the general soil biological community and these processes are analogous to that occurring with nitrogen. Moreover, (iii) the inter-conversion of mineral and organic nutrient forms with the quantity of water soluble phosphate is also controlled by dissolution rates from soil mineral fractions.

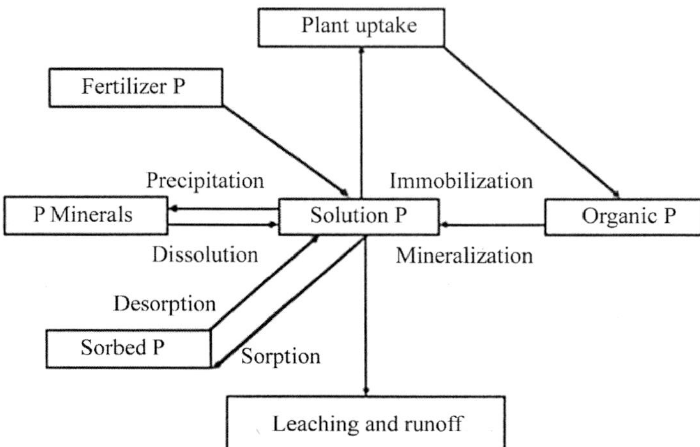

Fig. 1: Phosphorus cycle

In a broad sense, the P cycle in soil involves the uptake of P by plants and its return to the soil in plant and animal residues. Although much is known on the subject, specific information is required on dynamic processes occurring within the cycle, such as conversion of the P of plant and animal residues to inorganic phosphate through microbial activity, chemical and biochemical changes that continuously take place in the various P compounds of the soil, and transformations that occur as P is absorbed and utilized by microorganisms and higher plants. A complete understanding of the P cycle is of particular importance in tropical soils, where most of the P occurs in organic forms.

Unique properties of the phosphorus cycle includes the observation that phosphorus does not undergo any valence change in the cycle and that there is no gaseous component to the cycle. But P can exist in air either as phosphite (PO_3), phosphine (PH_3) and hydrogen phosphide (H_2P_4). Phosphorus in natural ecosystem is usually a scarce resource and is efficiently recycled. Approximately 0.1% of the estimated 10^{15} kg P in the earth's crust is subjected to at least some turnover. Estimates of total amounts of P occurring in terrestrial and oceanic reservoirs are:

	P reservoirs	Total P (x 10^{12} kg)
Land	Soil	96-160
	Mineable Rock	19
	Biota	2.6
	Fresh water (dissolved)	0.090

Ocean	Sediments	840,000
	Dissolved (inorganic)	80
	Detritus (particulate)	0.65
	Biota	0.050-0.12

Soil phosphorus compounds

Phosphorus compounds in soil can be placed into the following three classes: (1) organic compounds of the soil humus, (2) inorganic compounds in which the P is combined with Ca, Mg, Fe, Al, and with clay minerals, and (3) organic and inorganic P compounds associated with the cells of living matter, the soil biomass. Microorganisms are involved in transformations of P between organic and mineral forms. From 15 to 80% of the P in soils occurs in organic forms, the exact amount being dependent upon the nature of the soil and its composition (Anonymous). The higher percentages are typical of peats and uncultivated forest soils, although much of the P in tropical soils may occur in organic forms.

The chief source of organic P compounds entering the soil is the vast quantity of vegetation that undergoes decay. Recent evidence indicates that the readily decomposable or easily soluble fractions of soil organic phosphorus are often the most important factor in supplying phosphorus to plants. In contrast, the more soluble inorganic forms of phosphorus plays the biggest role in phosphorus fertility of less weathered soils, even though these generally contain relatively high amounts of soil organic matter.

A. Agricultural crops commonly contain 0.05 to 0.50% P in their tissues and in plants, this element is found in several compounds or groups of substances: phytin, phospholipids, nucleic acids, phosphorylated sugars, coenzymes and related compounds. Phosphorus may also present as inorganic orthophosphate, especially in vacuoles and internal buffers.

   The P in phytin, phospholipids and nucleic acids is found as phosphate. Phytin is the calcium magnesium salt of phytic acid, the latter being synonymous with inositol hexaphosphate. Inositol phosphates may have one, two, three, four, five or six phosphorus atoms per inositol unit and these several compounds have been found in soils and living organisms, and are formed enzymatically.

   Phospholipids are compounds in which phosphate is combined with a lipid. In a class of phospholipids that includes lecithin and cephalin, the phosphate is esterified with a nitrogenous base. For example, lecithin is made of glycerol, fatty acids, phosphate and choline.

The nucleic acids RNA and DNA consist of a number of purine and pyrimidine bases, pentose sugar and phosphate.

B. In bacterial cell, the bulk of P is in RNA, usually accounting for one-third to somewhat more than one-half of all the P. DNA contributes from 2 to 10% of the total P content. About 15-25% of total cell P content are found in acid-soluble organic and inorganic compounds. The acid-soluble fraction of bacterial protoplasm contains ortho- and metaphosphate, sugar phosphates, many of the coenzymes and adenosine phosphates. Inorganic polyphosphates may sometimes be quite abundant in certain fungi.

   Phospholipids usually represent less than 10% of the cell P and the concentration of phospholipids in bacteria varies greatly depending on species and age of the organism. Inositol phosphates may occur in microorganisms but there is no evidence for the presence of phytin.

C. In soil, from 15 to 85% of the total P is organic. The absolute quantity of organic P generally declines with increasing depth and the portion of total organic P is greater in surface than in subsurface horizons. Of the inorganic forms, large quantities occur in minerals where the phosphate is part of the mineral structure, as insoluble Ca, iron or aluminium phosphates. The Ca salts predominate in neutral or alkaline conditions whereas the Fe and Al salts predominate in acid surroundings.

   The organic compounds making up the humus fraction are derived from surface vegetation, microbial protoplasm, or metabolic products of the microflora. Of these substances, the inositol phosphates (10-80%), nucleic acids (1-10%), and phospholipids (0.1-5%) are significant components of soil organic fraction. The various inositol phosphates are often classified together as phytin or related substances. Such components frequently accounts for 10-80% of the entire organic phosphorus fraction. The molecules of inositol phosphates may have one to six phosphates and all of these exist in nature. The inositol hexaphosphate is dominant, the pentaphosphate is less abundant and the tetra, tri-, di- and monophosphates exist in still smaller quantities. It is believed that several of the inositol phosphates are of microbial origin because they are not known to occur in higher plants.

   The evidence for the existence of nucleic acids or nucleotide derivatives in soil is indirect. In most localities, the nucleic acid-type compounds probably contribute less than 1 to a maximum 10 per cent of the total organic phosphorus. As the microbial cell is rich in nucleic acids, it is possible that a reasonably large part of nucleic acid in soil (which is primarily of the RNA type) is bound within the cells of viable members of the microflora. Once the cells die, the nucleic acids would be readily mineralized.

The phospholipid content of humus is invariably small. Often 0.1 to 5.0 per cent or sometimes slightly more of the organic phosphorus is tied up in such compounds. A significant part of phospholipids may be phosphatidyl-ethanolamine and phosphatidylcholine and these compounds are found in both plants and microorganisms.

Soils rich in organic matter contain abundant organic phosphorus. Moreover, a good correlation exists between the concentrations of organic phosphorus, organic carbon and total nitrogen (almost all of this being organic). Ratios of organic carbon to organic phosphorus of 100 to 300:1 are common for mineral soils. Similarly, the nitrogen:organic phosphorus ratio may range from 5 to 20 parts of nitrogen for each part of phosphorus. The organic phosphorus level, therefore, is directly related to the concentration of other humus constituents, the phosphorus content being 0.3 to 1.0% and 5 to 20% of the carbon and nitrogen concentration, respectively. These results are consistent with the hypothesis that the relative constancy of the ratios in humus is governed by the composition and activities of the microflora.

Availability of phosphorus in Indian soils

The phosphorus nutrient is estimated to be in insufficient amounts in most of the Indian soils as available P. According to one of the compilation in 1994 (Anonymous) based on about 9.6 million soil tests for available P in Indian soils, it is observed that 49.3% of areas covering different states and Union Territories are in the low category, 48.8% in the medium category and 1.9% have high P status. The limited high P available areas were only in Assam, Himachal Pradesh and Rajasthan. Therefore, application of phosphatic fertilizers is unavoidable in intensive farming system.

Unlike nitrogen that comes from biological or industrial fixation, the source of P is only from phosphatic and S rocks, which are non-renewable source and use of P fertilizers lead to the depletion of these resources. The problem of P management in soil is very tricky. (i) Firstly, the P fertilizers applied or the phosphate ions present in the soil are often fixed or rendered unavailable under the most ideal conditions, (ii) secondly, depletion of natural resource and (iii) low grade phosphate rocks available in the country. Therefore, the import of phosphatic fertilizers is imperative which costs heavily.

Phosphorus availability

Under acidic conditions, P ions are present as H_2PO_4 but are subjected to fixation with hydroxides of Al and Fe at pH below 5.0. Near neutral pH, HPO_4^{2-} ions are usually present. But above pH 8.0, the PO_4^{3-} ions form $Ca_3(PO_4)_2$ and its availability is reduced drastically.

Solubilization of inorganic phosphorus

The insoluble phosphates in soil include tricalcium phosphate [$Ca_3(PO_4)_2$], oxi-apatites [$Ca_3(PO_4)_2.CaO$], hydroxy-apatites [$Ca_3(PO_4)_2.Ca(OH)_2$], carbonate-apatites [$Ca_3(PO_4)_2.CaCO_3$] and fluor-apatites [$Ca_3(PO_4)_2.CaF_2$] in saline and saline alkaline soils and hydroxyl-phosphates of Fe and Al namely dufrenite, strengite, variscite etc. in acidic soils.

These insoluble inorganic compounds of phosphorus are largely unavailable to plants, but a large number of heterotrophic and autotrophic soil microorganisms have the capacity to solubilize inorganic phosphates through their metabolic activities directly or indirectly. About 10-50% of the bacterial isolates tested are capable of solubilizing calcium phosphates and counts of bacteria solubilizing insoluble phosphates may range from 10^5 to 10^7 per gram.

Bacteria belonging to diverse groups from autotrophs to heterotrophs, diazotrophs to phototrophs, fungi including mycorrhizal fungi, both ectotrophic as well as endotrophic and actinomycetes have been identified as active phosphate solubilizers. The major group of bacteria, fungi and actinomycetes identified as active phosphate solubilizers are listed.

Bacillus sp. - *B. circulans, B. megaterium, B. mycoides, B. mesentricus, B. megaterium var. phosphaticum, B. polymyxa, B. subtilis* and *B. coagulans*.

Pseudomonas sp. - *P. fluorescens, P. putida, P. striata, P. liquifaciens, P. rathonis.*

Species of *Flavobacterium, Brevibacterium, Alcaligens, Achromobacter, Aerobacter aerogenes, Xanthomonas, Erwinia, Serratia, Micrococcus, Escherichia freundii, E. intermedia, Nitrosomonas, Nitrobacter, Ferrobacillus ferroxidans, Thiobacillus ferroxidans, T. thioxidans* and *R. meliloti.*

Fungi

Aspergillus spp. - *A. flavus, A. awamorii, A. niger, A. terreus, A. fumigatus, A. nidulans*

Penicillium spp. - *P. lilacinum, P. digitatum*

Species of *Fusarium oxysporum, Trichoderma viride, Curvularia lunata, Sclerotium rolfsii, Alternaria teneuis, humicola, pythium, Phoma, Acrothecium, Morteirella, Paecilomyces, Cladosporium, Rhizoctonia, Rhodotorula, Candida* sp., *Cunninghamella, Oideodendron, Pseudogymnoascus.*

Actinomycetes

Actinomyces, Streptomyces.

Isolation of phosphate solubilizing microorganisms

The isolation of phosphate solubilizing microorganisms is mainly done by using insoluble phosphates like tricalcium phosphate, apatite, rock phosphate and in some cases, Fe and Al phosphates in agar media and observing clearing zones around colonical growth of microorganisms. These microorganisms not only assimilate the element but they also make a large portion soluble, releasing quantities in excess of their own nutritional demands. The solubilization of different types of insoluble phosphates varies with the type of microorganisms, the type of phosphates available, media conditions and available carbon source.

Mechanism of solubilization

(i) The dissolution of rock phosphates or insoluble phosphorus compounds by different microorganisms is mainly by production of organic acids in the medium. In the special case of the ammonium and sulphur oxidizing chemoautotrophs, nitric acid and sulfuric acids (inorganic acids) are responsible. The organic and inorganic acids convert $Ca_3(PO_4)_2$ to di- and monobasic phosphates with the net result of an enhanced availability of the element to plants.

Nitric and sulfuric acids produced during the oxidation of nitrogenous materials or inorganic compounds of sulfur react with rock phosphate and thereby increase the soluble phosphates. The oxidation of elemental sulfur is a simple and effective means of providing utilizable phosphates. For example, a mixture may be prepared with soil or manure, elemental sulfur, and rock phosphate, as the sulfur is oxidized to sulfuric acid by *Thiobacillus*, there is a parallel increase in acidity and net release of soluble phosphate.

Nitrification of ammonium salts also leads to a slight but significant liberation of soluble phosphorus from rock phosphate composts. Biological sulfur or ammonium oxidation has never been adopted on a commercial scale because of the availability of cheaper and more efficient means of preparing fertilizers.

(ii) Although phosphate solubilization commonly requires acid production, other mechanisms may account for ferric phosphate mobilization. In flooded soil, the iron in insoluble ferric phosphates may be reduced leading to the formation of soluble iron with concomitant release of phosphorus into solution. Such increases in the availability of phosphorus on flooding may explain why rice cultivated under water has a lower requirement for fertilizer phosphorus than the same crop grown in dry land agriculture.

(iii) Phosphorus may also be made more available for plant uptake by certain bacteria that liberates hydrogen sulfide, a product that reacts with ferric phosphate to yield ferrous sulfide and liberates the phosphate. Fermentative microorganisms produce hydrogen sulfide from sulfur-containing amino acids or anaerobic sulphate reducing bacteria like *Desulphovibrio* and *Desulfatomaculum* causes reduction of sulfate to hydrogen sulfide when the redox potential is low.

(iv) The CO_2 produced in the rhizosphere due to decomposition of organic matter by microorganisms has also been reported to be involved in increased phosphorus availability to plants. The reaction may be with CO_2 directly or due to formation of carbonic acid which react with $Ca_3(PO_4)_2$ forming $CaHPO_4$ or $Ca(H_2PO_4)_2$ and $CaCO_3$.

Factors affecting solubilization

(i) The amount brought into solution by heterotrophs varies with the carbohydrate oxidized, and the transformation generally proceeds only if the carbonaceous substrate is converted to organic acids. Under cultural conditions, bacteria are more active in the presence of hexoses and pentoses in the medium whereas fungi are equally effective in the presence of hexoses, pentoses as well as disaccharides.

(ii) The quality of rock phosphate and its particle size also affect solubilization. Normally high-grade rock phosphates, with fine particle size are dissolved readily than low grade phosphates or coarse size particles.

(iii) Fungi normally are active at a wide pH range from acidic to neutral whereas bacteria are more active above neutral pH.

(iv) Aeration enhances solubilization and the rate of solubilization is maximum between 25 and 30^{0}C temperature.

Solubilization in situ in plant rhizosphere

The phosphate-dissolving microorganisms are generally found in greater numbers in plant rhizosphere than non-rhizosphere soil. Because, the plant root exudates provide readily metabolizable carbon and nitrogen compounds for the growth of heterotrophic forms. The bacteria and fungi form an associative symbiosis with the root system to get substrates from the roots and in return provide mineral nutrients which otherwise normally could not be absorbed by the roots including phosphorus.

The higher populations of bacteria and fungi capable of dissolving insoluble phosphates in rhizosphere as compared to non-rhizosphere have been observed in the rhizosphere of wheat and maize, oats, rye grass, mustard, rice and jute, and legumes such as berseem, green gram, lupins, cowpea, soybean, groundnut, black gram and chickpea. These microbial count studies of phosphate solubilizing bacteria (PSB) indicate that because of greater populations of phosphate solubilizing microorganisms in the rhizosphere, the phosphate availability to plants is increased.

Seed bacterization with phosphate solubilizing microorganisms

Phosphatic biofertilizers were first prepared in USSR using *Bacillus megaterium* var. *phosphaticum* as phosphate-solubilizing baacteria and the product was named as 'phosphobacterin' which was extensively used in collective farming for seed and soil inoculation to cover an area of 14 million hectares annually and reported to give 5-10% increase in crop yields.

Inoculation experiments conducted with phosphobacterin and other phosphate-solubilizing microorganisms in various crops like oat, wheat, potatoes, peas, tomatoes and tobacco showed an average 10-15% increase in crop yields in about 30% of the experiments conducted. Several experiments conducted in legume and non-legume crops by co-inoculation of PSM with diazotrophs have shown synergistic effects with regard to increase in population of both bacteria and significant increase in crop yields in comparison to single inoculation.

Mineralization of organic phosphorus

In this process, microorganisms convert the organic phosphorus to inorganic forms. Thus, the bound element in the plant residue material and in soil organic matter is made available to succeeding populations of plants by the actions of bacteria, fungi and actinomycetes.

The mineralization and immobilization of this element are related to the analogous reactions of nitrogen. As a rule, phosphate release is most rapid under conditions favouring ammonification (N mineralization). Thus, a highly significant correlation is observed between the rates of nitrogen and phosphorus conversion to inorganic forms. The nitrogen mineralized being from 8-15 times the amount of phosphate made available. There is also a correlation between C (CO_2 release) and phosphorus mineralization i.e. a ratio of 100 to 300:1. These results show that the ratio of C:N:P mineralized microbiologically at the equilibrium conditions is similar to the ratios of three elements in humus.

The enzymes that cleave phosphorus from the more frequently encountered organic substrates are collectively called phosphatases. These enzymes catalyze the following reactions:

A single phosphatase enzyme may catalyze the cleavage of ethyl phosphate, glycerophosphate and phenyl phosphate that is R may have numerous structures. On the other hand, molecules with two R groups (known as diesters, in contrast with monoesters having only one) may require different enzymes for their breakdown. Phosphatases acting on phospholipids and hydrolyzing nucleic acids have diesters as their substrates. The phosphatase enzyme catalyzing hydrolysis of the monoesters often has distinct optima in pH for maximum activity i.e., active at low pH ranges - acid phosphatases; and active at high pH ranges - alkaline phosphatases. The abundance of organisms in soil, possessing these enzymes is shown in table.

Substrate	Per cent isolates able to cleave substrate	
	In soil	In soil near roots
Glycerophosphate	22-26	18-36
Lecithin	11-20	12-38
Phytin	30	42
Phenolphthalein diphosphate	70-74	58-76

The enzyme phytase liberates phosphates from phytic acid or its Ca-Mg salt, phytin with the accumulation of inositol. The enzyme acts on the hexaphosphates and remove the phosphate, presumably one at a time to yield the penta-, tetra-, tri-, di- and monophosphates and then finally free inositol.

Phytase enzyme characteristics

(i) Some organisms cleave only the penta- or tetra- but not the hexaphosphates. (ii) Some species make intracellular phytase while others excrete extracellular phytase enzyme. (iii) Moreover, some phytases are reasonably specific and act chiefly or solely on inositol phosphates. Others are really nonspecific phosphatases removing phosphorus from dissimilar organic compounds.

Phytase activity is widespread and about 30-50% of the isolates from soil synthesize this enzyme and its activity in nature is enhanced by addition of carbonaceous materials that increase the size of community. Species of Aspergillus, Rhizopus, Cunninghamella, Arthrobacter, Streptomyces, Pseudomonas and Bacillus have been found to synthesize the enzyme.

DNA or RNA are large complex molecules and cannot be transported the cell membrane. Pure nucleic acids added to soil are rapidly dephosphorylated. Indeed, a large number of different heterotrophs can grow in media containing nucleotides as the sole sources of carbon, nitrogen and phosphorus. The transformation proceeds by an initial depolymerization of RNA by ribonuclease and of DNA by deoxyribonuclease and subsequent cleavage of the phosphate from the products generated by depolymerizing enzymes (probably by the phosphatases). The mineralization of phosphate from nucleic acids is affected by pH and the rate declines as the acidity increases.

Bacteria, fungi and actinomycetes also cleave phospholipids for use as a phosphorus source. Lecithin is the common substrate for the assessment of such actions. In this mineralization process, phosphate is cleaved from the organic compound and is then assimilated by the responsible populations. If the phosphate release exceeds the demand of heterotrophic community, some phosphate will be available for plant uptake also.

Measurement of phosphatase activity

The phosphatase activity in many soils has been measured. For this purpose, a soil sample is incubated with a substrate such as glycerophosphate or β-naphthyl phosphate, phenyl phosphate or another organic compound, and the liberation of either inorganic phosphate or the organic portion of the substrate (e.g. β-naphthol or phenol) is determined. These assays show that the enzymatic activity is usually high and is affected by season, depth and type of vegetation. The hydrolysis occurs at low and high pH, a possible indication of both acid and alkaline phosphatases.

Factors affecting mineralization of phosphorus

- Mineralization is usually more rapid in virgin soils (more organic carbon, more organic phosphorus, more microbial population, more mineralization) than in the cultivated soils.
- The decomposition is also favoured by warm temperatures, with the thermophilic range being more favourable than the mesophilic range.

- The rate of mineralization is enhanced by adjusting the pH values conducive to general microbial metabolism and a shift from acidity to neutrality increases phosphate release.
- The rate of mineralization is directly correlated with the quantity of substrate, hence soils rich in organic phosphorus will be the more active.
- The mineralization is not inhibited by inorganic phosphorus and the mineralization proceeds rapidly even at sites having adequate phosphate.

As expected, phosphorus uptake by plants is correlated with the mineralization rate.

Immobilization of phosphorus

Microbial growth requires the presence of available forms of phosphorus, because the phosphorus element is essential for cell synthesis. Therefore, the development of the microflora is governed by the quantity of utilizable phosphorus compounds in the habitat. In environments, where phosphorus is limiting, its addition will stimulate microbiological activities. Deficiencies may be induced artificially by the addition of carbohydrates. The added carbohydrates will have stimulatory effect on proliferation of microorganisms that will require additional amount of C, N and P for immobilization purposes. Therefore, it will lead to phosphorus deficiency in the surrounding ecosystem.

The assimilation of phosphorus into microbial nucleic acids, phospholipids and other protoplasmic substances of microbial cells leads to the accumulation of non-utilizable forms of the element (because utilizable form has been assimilated). Hence, during the decomposition of organic matter added to soil, the increase in microbial abundance puts a great demand on the phosphate supply and if the added carbonaceous residue is deficient in phosphorus, the microbial assimilation of available phosphate may depress crop yields. Such decreases in yield can be prevented by the application of phosphatic fertilizers.

The microbial or plant utilization of phosphate can be measured by comparing the growth and chemical composition of plants grown in the presence and absence of microorganisms. Thus, when barley is seeded into sterile and nonsterile samples of soil with a low level of available phosphate, the yield and phosphorus content of plants are less in the nonsterile samples (Table). This suggests that the subterranean community is competing with the plants. The reduction in growth by the microflora in natural soil disappears if the phosphate supply is high.

Response measured	Plant part	Quantity per plant (mg)	
		Sterile soil	Non-sterile soil
Dry matter	Shoot	370	240
	Root	240	120
P content	Shoot	0.35	0.23
	Root	0.14	0.06

Phosphorus, like nitrogen is, therefore, both mineralized and immobilized. The process that predominates is governed by the percentage of phosphorus in the plant residues undergoing decay. If the phosphorus concentration exceeds than that required for microbial nutrition, the excess appears as inorganic phosphate. If phosphorus concentration is inadequate for the microflora, the net effect results in immobilization. Thus, in the decomposition of substrates poor in phosphorus which have a wide C:P ratio, a portion of the available nutrient supply is immobilized from the surroundings. As the ratio (C:P) narrows with time because of CO_2 volatilization, phosphate will accumulate.

Consider a hypothetical case of organic matter decomposition in which 100 parts of carbonaceous material containing 40 per cent carbon are acted on by fungal, bacterial and actinomycetes population. The mineral composition studies of bacterial and fungal cells showed that (a) commonly phosphorus accounts for 0.5 to 1.0% of fungus mycelium and (b) 1.0 to 3.0% of the dry weight of bacteria and probably actinomycetes.

Assuming further that 30-40% of substrate carbon is assimilated during decomposition by fungi, 5-10% of substrate carbon is assimilated by bacteria and 15-30% of substrate carbon is assimilated during decomposition by actinomycetes, then the following calculations present a first approximation to the critical phosphorus content of organic materials.

(i) Substrate carbon added = 100 parts of 40% C (100x0.40 = 40 parts of substrate C added).

(ii) Fungi: 30-40% substrate C is assimilated i.e. 0.30 to 0.40 of 1 part C. Thus (0.30-0.40)x40 parts substrate c assimilated = 12 to 16 parts substrate cell C formed.

(iii) Fungi commonly contain 50% C (50% other material). Hence fungal mass formed = (12 to 16)/0.50 = 24 to 32 parts of fungal mass formed.

(iv) Fungal mycelium contains 0.5 to 1.0% phosphorus i.e. 0.005 to 0.01. Therefore, P required per 100 parts of organic matter = (0.005 to 0.01) x (24 to 32) = 0.12 to 0.32 parts.

Similar calculations give values of 0.06 to 0.20 parts (by bacteria) and 0.18 to 0.60 parts (by actinomycetes) of the phosphorus needed for the decomposition of 100 parts of substrate. In actual trials in media containing glucose as the carbon source, *Aspergillus niger* assimilates 0.24 to 0.40 parts, *Streptomyces sp.* assimilates 0.27 to 0.63 parts and mixed flora range from 0.16 to 0.36 parts of phosphorus for each 100 parts of glucose oxidised.

With a carbohydrate like cellulose, 0.35 to 0.45 parts of phosphorus are assimilated by the microflora for each 100 parts of cellulose. Similarly, in a soil receiving a continuous supply of glucose, 0.37 parts of phosphorus are removed by the community for every 100 parts of the sugar.

A value of 0.3 may be taken as an average figure for aerobic conditions, provided the substrate is readily and completely metabolized. Anaerobically, microorganisms derive less energy from decomposition and consequently fewer cells are synthesized, hence less phosphate need to be metabolized for the same quantity of fermentable carbon.

In the decomposition of natural substrates (as compared to glucose, cellulose or other sugars utilization), the quantity of phosphorus immobilized is diminished because less of the total carbon is degraded in a finite (unit) time interval. As a consequence, the amount of phosphorus immobilized during the decomsposition of plant residues is closer to 0.2% of the dry weight of the organic matter. Therefore, the critical level of phosphorus in natural carbonaceous products that serves as a balance point between immobilization and mineralization is 0.2%. If the substrate contains more phosphorus than the critical level, some phosphorus is released. If the material has less phosphorus than the critical level (less than needed by the microflora), phosphorus disappears from the environment, as the net effect is immobilization. On the other hand, as the phosphorus poor organic matter is decomposed and the phosphorus content of the residue increases, a point is eventually reached when the 0.2% figure is exceeded, and phosphate reappears.

Objective type questions

A. Fill up the blanks with suitable words

i. ______________________ is an example of vesicular arbuscular mycorrhizae (VAM).

ii. The nutrient element _________________ is known as energy currency of plants.

iii. Agricultural crops commonly contain _______ to _______ % P in their tissues

iv. The highly unavailable form of phosphorus in alkaline soil is ___________.

v. _______________ and _______________ are the examples of phosphorus solubilising bacteria.

vi. _______________ and _______________ are the examples of phosphorus solubilising fungi.

vii. _______________ and _______________ are the examples of phosphorus solubilising actinomycetes.

viii. The enzymes that cleave phosphorus from the more frequently encountered organic substrates are collectively called _______________.

B. Multiple choice questions

i. Which of the following nutrient regulates root and shoot growth -

a. P b. N

c. K d. Zn

ii. Which of the following nutrient is known as energy currency of plants

a. P b. N

c. S d. Zn

iii. In soil, organic phosphorus content generally varies between -

a. 15-85% b. 25-90%

c. 80-90% d. None

iv. Ratios of organic carbon to organic phosphorus of _______________ are common for mineral soils

a. 50 to 100:1 b. **100 to 300:1**

c. 300 to 400:1 d. None

v. The highly water soluble form of phosphorus in soil solution is -

a) $H_2PO_4^-$ b) HPO_4^{2-}

c) PO_4^{3-} d) All are equal

vi. The phosphate solubilising bacteria are generally _____________ in their nutrition.

a) Heterotrophic
b) Autotrophic
c) Phototrophic
d) Chemoautrophic

vii. Phosphorous solubilisation in flooded soil is due to the presence of

a. Carbonic acid
b. Sulfuric acid
c. Nitric acid
d. All of these

viii. Immobilization of inorganic phosphorus occurs at the C:P ratio in soil is -

a. >100:1
b. >200:1
c. >300:1
d. None

Answer keys

A. Fill in the blanks

i. *Glomus*, **ii.** Phosphorus, **iii.** 0.05 to 0.50% **iv.** Apatite, **v.** *Bacillus* and *Pseudomonas*, **vi.** *Fusarium* and *Aspergillus*, **vii.** *Actinomyces* and *Streptomyces*, viii. Phosphatases.

B. Multiple choice questions

i. **a** ii. **a** iii. **a** iv. **b** v. **a** vi. **a** vii. **d** viii. **c**

C. Descriptive type questions

i. Enlist in brief the different forms of soil phosphorus. Write two examples each of phosphate solubilising bacteria, fungi and actinomycetes.

ii. What do you mean by phosphorus mineralization-immobilization reaction in soil? Briefly describe.

iii. What are the factors affecting microbial transformation of soil phosphorus?

iv. Describe phosphorus cycle with suitable pictorial diagram.

14

Microbial Transformation of Sulfur in Soil

Introduction

In soil, microorganisms influence not only the availability of carbon and nitrogen, which are important constituents of plants, but also of sulfur (S), phosphorus and many other trace elements for plant absorption. Sulphur is found widely distributed and considered as tenth most abundant element in the nature. Sulfur is considered as an essential element for survival of plant and animals and found as gypsum ($CaSO_4.2H_2O$), pyrite (FeS_2) and other combined form and also in elemental form (S^0) in the nature. The requirement of sulphur by plant is comparatively less than the primary nutrients, hence sulfur is recognised as 'secondary' nutrient. The addition of S in fields in the past was not given due priority mainly because of limited area or crops that gave response with the added fertilizer and addition through the major fertilizers and natural sources. Addition of S in soil with pesticides, fertilizer, irrigation water and adsorption in the form of SO_2 gases from atmosphere is significant. Alone rainfall can contribute 5 to 250 kg/ha/year of S in soil depending on fossil fuel burning and industrial activity. Therefore, S deficiency generally observed in soils away from industrial activity and sea. The lithosphere is found to be the major sink of sulfur (24.3×10^{18} kg) followed by hydrosphere (1.3×10^{18} kg), pedosphere (2.7×10^{14} kg) and atmosphere (4.8×10^{9} kg), in the earth. The atomic weight of S 32.064 and it exist in a number of oxidation states, as indicated from the different oxidation number in several compounds viz. sulfides (-2), polysulfide (-1), elemental sulfur (0), thiosulfate [(-2) and (+6)], sulfite (+4) and sulfate (+6).

Source of sulphur in soil and its different pools

Soil organic matter is considered as the major S source in majority of the cases. Organic form of sulphur is found around 75-90% of total amount in soil, whereas, only 10-15% of S is present in the inorganic form (sulphate). Thus, a minor portion of the total sulfur content is present in soil as inorganic component. Due to the weathering sulfur released from the mineral as sulfate (SO_4^{2-}) which is uptake by plants and converted to different organic forms (Fig. 1). Soil microorganisms metabolized the organic form of S when added in soil in bulk and converted to inorganic state (sulfur, sulfates, thiosulfate, sulfite etc.) for plant uptake and a minor amount is converted to humus. Sulfur in organic state, is bound in proteins of plant and animal origin and also in microorganism in the protoplasm as sulfur containing amino acids (cysteine, cystine, methionine), proteins, lipid, polypeptides, thiamine, biotin etc. The organic sulfur compounds thus can mainly be divided in ester sulfates group, with C-O-SO_3 linkages; and carbon-bonded S group, with direct C-S linkages. Other existed organic forms of S are of minor importance. Ester sulfates comprised compounds like phenolic sulfates, choline sulfate and sulfated polysaccharides. The carbon-bonded S is majorly includes amino acids like cysteine and methionine, and sulpholipids. As compared to the C-bonded S, ester sulfate mineralized faster and acts as readily available source of S for plants as well as microbes. Organic form of S may present in soil more than 95% of the total amount of S in temperate condition. In arid or semi-arid region, appreciable amount of inorganic sulfur also present in soil. In soil, minerals that contribute sulfur are mainly pyrites (FeS_2), chalcopyrites ($CuFeS_2$), sphalerite (ZnS), epsomite ($MgSO_{4,}6H_2O$), gypsum, etc. As compared to the coarse texture soil more sulfur is present in fine texture soil and in lower depth of soil as compared to the surface due to difference in organic carbon distribution. Due to the weathering process S release from mineral as sulfate (SO_4^{2-}) and consumed by plants and then converted to different organic form (Fig. 1). After addition of organic form of S to the soil in bulk, it is metabolized by microorganisms and contribute to the plant nutrition, significantly.

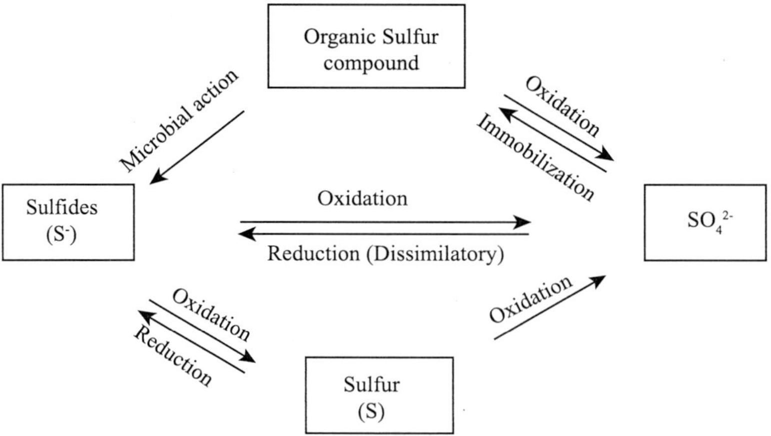

Fig. 1: Role of microorganism in Sulfur cycle

Functions of sulfur

In healthy plants, S ranges from 0.1 to 0.4% based on dry weight. Sulfur play important role in several functions in plant. Ferrodoxin, the first stable compound of photosynthetic electron chain is a Fe-S protein. Sulfur is involved in the glucosides synthesis in mustard oil. Sulfur is constituent of several components in plants viz. glutathione, thiamine, coenzyme A, amino acids (cysteine, cystine and methionine) etc. Uptake of sulfur for grain production was highest in oilseed crops followed by pulses and cereals, respectively. Deficiency of S results decrease in photosynthetic activity, retardation of growth, yellowing of mainly young leaves and ultimately poor crop yield. Nitrogen fixation is also retarded in S deficiency as both are component of protein. Thus for optimum N metabolism and ultimately for obtaining good yield desired N:S ratio should be maintained. The optimum N:S ratio for legume is around 15-16:1 and for cereals the ratio is 11-12:1.

With high analysis fertilizer application, increased use of hybrid or high yielding cultivar and exhaustive cropping system, sulfur deficiency is reported from larger area and variety of crops. Sulfur availability of plants largely depend on dynamic sulfur cycle which is influenced largely by the native microbial biomass in the soils. Especially in the soil, microbes play a pivotal role in rhizosphere, allowing plants for uptake of organo-sulfur by the process of minerelization and subsequently conversion to inorganic plant available form. This chapter will deals with the soil microbial processes involve in sulfur cycling and microbial community that influence these processes.

Cycling of sulfur in soils

Sulfur cycle in soil is much similar to that of nitrogen. Cycling of sulfur between inorganic and organic states and between reduced and oxidized states are operated by various microorganisms, majorly bacteria. Thus mineralization of sulphur refers the transformation of organically bound sulfur to the inorganic state by soil microorganisms. The sulfur / sulfate, thus released are absorbed either by the plants or releases to the atmosphere as oxides. In case of insufficient oxygen, hydrogen sulphide is produced by certain group of microorganisms from organic S compounds, particularly in waterlogged soils. Several studies both chemical and spectroscopic have confirmed that most of the soil sulfur (>95%) in agricultural soils is existed the form of sulfate esters or as carbon-bonded sulfur (sulphonates or amino acid sulfur), instead of inorganic sulfate. Plant primarily consumed inorganic sulfate. Although, findings from recent study has elaborated that the sulphonate and sulfate ester – pool of soil S can also be bioavailable for plant, might be due to transformation of carbon-bonded and sulfate ester sulfur to inorganic sulfate by the action of soil microbes. However, in addition to the mineralization of bound forms, soil microbes are responsible also for the immobilization of sulphate to first sulfate esters and then to carbon-bound sulphur. Microbial community present in soil govern the rate of sulfur cycling although yet it is not clear, whether a particular microbial species or genera control this process or not. A common rhizosphere bacterium, *Pseudomonas putida*, involved in the mobilization of sulphonate- and sulfate ester sulphur. Some mutants of this species show reduced survival in the soil, as that are not able to transform sulfate esters reflecting their dependency on sulfate ester compounds for S nutrition. Moreover, it had been studied that mutants that are unable to metabolize sulphonate-sulfur viz. *P. putida* S-313 ,do not support the growth of tomato crop as the wild-type strain does, indicating the role of mobilize sulfur for plant nutrition.

Therefore, the S cycle in soils is dynamic in nature. In the S cycle, inorganic compounds are immobilized to organic compound of sulfur, different organic compounds transformed, and immobilized S is subsequently mineralized to give inorganic plant available sulfur. Most of these inter-conversion process are linked with the presence of microbial biomass in the soils. Microbes, especially in the root rhizosphere, play a important role in facilitating plants to consume soil organosulfur. This chapter will discuss the microbial processes in soil that influence sulfur cycling, and will highlight the micro-organisms that catalyse these transformation processes.

Transformations of sulfur in soil

Sulfur transformations in the biosphere can be summerize as a cyclic process including: (1) decomposition of organic sulfur to inorganic compounds via many subunits, through the mineralization process, (2) immobilization process where, sulfur assimilation was done into the protoplasm of microorganisms, (3) oxidation process where inorganic sulfur converts into elemental sulfur, and (4) sulfate reduction. Microorganisms both aerobic and anaerobic are involved in formation of organic S, - although S constituents only 1 to 3 percent of microbial biomass. However, the short life span of microorganisms, result in rapid S recycling. Microbial biomass is considered as the most active and readily available pool of organic form of S in soil, and most of the mineralized S, found in short-term incubation experiments may come from microbial biomass.

Much of the S transformations in soil are results of microbial activity, though chemical transformations are also found in some cases (e.g. oxidation of iron sulphide). The major types of transformations in the S cycle are: mineralization, immobilization, oxidation and reduction.

Sulfur mineralization in soil

Sulfur is consumed by the root system of plant, mostly as the sulfate ion although many amino acids, without prior degradation may also be assimilated. Sulfur is essential for growth of agricultural crops and other vegetation and as a result sulfate is found in their rooting medium. The mineralization of organic sulfur to ultimately sulphate plays an important role in the microbiological reactions required for higher life. Organic S compounds breakdown to smaller units and subsequently converted into inorganic compounds (sulfates) by the soil microorganisms. Sulfur mineralization rate is about 1.0 to 10.0 per cent/ year. Various organic S compounds are act as substrates to the microbes. The S present in plant, animal and microbial proteins in the form of amino acids, cystine and methionine, and as vitamin B, biotin, thiamine, and thioctic acid. It is also present in the animal tissues and excretory products as free sulfate, taurine and, as thiosulfate and thiocyanate, to some extent.

Addition of plant and animal remains to soil results the minerelization of S contained compound. Microbes utilize a minor portion of the inorganic products for cell synthesis, and the rest is released to the environment. In aerobic condition, the inorganic, terminal product is sulfate. In absence of atmospheric O_2, specifically during the putrefaction of protein compound, H_2S and the odoriferous mercaptans persists. There are a number of bacteria genera able to form H_2S from partially degraded proteins. Therefore, sulfides are among the major inorganic compounds formed during the decomposition of proteinaceous substrates. When cystine or cysteine is applied to an aerated soil, the sulfur in these amino acids is recovered quantitatively as sulphate.

Decomposition may occur by any of several known mechanisms. Chemical oxidation of added cysteine in soil can form cystine. Upon oxidation of the sulfur in the molecule, cystine disulfoxide and possibly cysteine sulfinic acid are produced as intermediates, a reaction sequence that does not involve H_2S.

Several microorganisms attack the two compounds, which results in rapid conversion. Some fungi, like *Microsporeum gypseum*, convert cysteine-sulfur to sulfate via a process in which cystine, sulfenic and sulfinic acids, sulfite, and sulfate are formed sequentially. Similar to the ammonification of organic nitrogen, the amount of mineral sulfur formed depends on the sulfur content and the C:S ratio of the decomposers. When the organic matter contains more sulfur than is needed by the microbes, sulfate accumulates.As there are no precise data, it is likely that, in humid-temperate soils, about 1 to 3% of the total sulfur supply is mineralized per annum, similar to nitrogen. The rate of sulfur mineralization could also be affected by environmental factors that regulate microbial growth in general.

Immobilization of sulfur

During immobilization of sulfur, inorganic sulfur compounds are converted to organic sulfur compounds by microbes. Microbes use many different compounds as sulfur sources for its growth, but any one strain may depend on only a few substances. Among the inorganic substances that may provide this element are sulfate, thiosulfate, hyposulfite, sulfoxylate, persulfate, sulfide, elemental sulfur, tetrathionate, sulfite, and thiocyanate, and the organic substances are cystine, cysteine, methionine, taurine, and undecomposed proteins. In soil anaerobes probably assimilate reduced sulfur compounds, since sulfur is not produced in an environment devoid of oxygen. Both aerobic and anaerobic chemotrophs and phototrophs perform sulfate immobilization as a reductive process. In contrast, some anaerobic microbes (e.g. phototrophic green S bacteria) are just capable of sulfide immobilization for their needs, which is less energy-consuming compared to sulfate assimilation. Microbial tissue has a C/S ratio between 57 and 85 in bacteria, and 180 to 230 in fungi. Most microorganisms have a sulfur content between 0.1 and 1.0% of the dry weight, with cystine and methionine being the most conspicuous cellular constituents containing sulfur.Sulfur immobilized in organic matter by covalent bonding. When added to soil, sulfate can adsorb quickly or be transformed into low molecular weight organic sulfur compounds, especially ester sulfates as fulvic acids, which can later be polymerized to a larger amount of insoluble organic compounds. Measuring the inorganic sulphate released in chloroform fumigation technique, it is possible to estimate how much sulfur is immobilized by microbes. Despite the fact that there is less sulfur sequestered in the microbial biomass, the fraction is very labile and an important indicator of plant availability. Sulfate application can prevents crop yield reductions caused by starch addition to sulfur-poor soils. As a result of microbial utilization of the available sulfur during the starch decomposition, the nutrient becomes immobilized leading to the detrimental effect. It is reported that the C:S ratio in carbonaceous materials above which immobilization dominates mineralization is approximately 50:1..Diversity within soil microbes, critical C: S ratio of the substrate, and environmental factors viz. moisture, temperature, atmospheric deposition inputs, organic matter, and other factors were known to influence immobilization rates.

Oxidation of inorganic sulfur

Chemoautotrophous and photosynthetic bacteria are capable of oxidizing sulfur compounds (such as hydrogen sulfide, sulphite, and thiosulfate) to sulfate (SO_4).Plant and animal protein degradation releases sulfur which accumulates in the soil, which, in the presence of oxygen, is oxidized into sulfates, whereas in anaerobic conditions (waterlogged soils), organic sulfur decompose to release hydrogen sulfide (H_2S).Under anaerobic conditions, reduction of sulphates may cause accumulation of H_2S, which again under aerobic conditions, could oxidized to sulphates

a) $2S + 3O_2 + 2H_2O$ --------> $2H_2SO_4$ --------------> $2H^+ + SO_4$ (Aerobic) (Ionization)

b) $CO_2 + 2H_2S$ --------------> $(CH_2O) + H_2O + 2S$ (Light)

OR

$H_2 + S + 2CO_2 + H_2O$ ---------> $H_2SO_4 + 2(CH_2O)$ (Anaerobic) (Light)

Bacteria of genus *Thiobacillus*

The main organisms responsible for oxidation of elemental sulfur to sulphates are the members of genus *Thiobacillus* (obligate chemolithotrophic, non photosynthetic) e.g., *T. ferrooxidans* and *T. thiooxidans* In this case, these are aerobic, non-filamentous, chemosynthetic autotrophs Green and purple sulfur bacteria.

In aquatic environments, several green and purple bacteria (Photolithotrophs) from the genus *Chlorbium, Chromatium*, and *Rhodopseudomonas* are also known to oxidize sulfur. They are members of the families Thiorhodaceae and Chlorohacteriaceae. Green and purple bacteria, which developed anaerobically, obtain their energy from light, and oxidize reduced sulfur materials, with CO_2 as their only carbon source. In bottom of water bodies containing sulfur material, they are more commonly found.

Colourless filamentous sulfur bacteria

Colourless filamentous sulfur bacteria of the Species *Thiothrix*, *Beggiatoa*, *Thioploca* and *Thiospirillopsis* are generally found in sulfide-containing waters and able to oxidize sulfide to sulphate resulting accumulation of elemental sulfur in cells. In addition, heterotrophic bacteria such as *Bacillus, Pseudomonas*, and *Arthrobacter*, as well as fungi (*Aspergillus, Penicillium*) and some actinomycetes have also been found to oxidize sulfur compounds. During oxidation of sulfur and H_2S, sulfuric acid forms that is of great significance for reducing the pH of alkaline soils and preventing potato scabs and rots caused by *Streptomyces* bacteria. The formation of sulfate / sulfuric acid is advantageous to agriculture in several ways: (i) it makes alkali soils fit for cultivation by reducing soil pH; (ii) increasing the levels of phosphates, potassiums, calciums, magnesium etc. for plant nutrition by solubilizing inorganic salts containing those plant nutrients.

Reduction of sulfate

Sulfate present in the soil is assimilated and incorporated into proteins by plants and microorganisms known as "assimilatory sulfate reduction". Sulfate reducing bacteria (e.g. *Desulfovibrio* and *Desulfatomaculum*) can reduce sulfate to hydrogen sulphide (H_2S) and may decrease the sulfur available for plant nutrition, known as "dissimilatory sulfate reduction" which negatively affected soil fertility and agricultural productivity.

Alkaline and anaerobic soil conditions are conducive to dissimilatory sulfate-reduction, causing sulfates to be reduced to hydrogen sulfide. Calcium sulfate, for example, is attacked by members of the genus *Desulfovibrio* and *Desulfatomaculum* under anaerobic conditions to release H_2S.

$$CaSO_4 + 4H_2 \xrightarrow{\text{Light}} Ca(OH)_2 + H_2S + H_2O$$

Several species of green and purple phototrophic bacteria (e.g. *Chlorobium, Chromatium*) further oxidize hydrogen sulphide produced by the reduction of sulfate and sulfur-containing amino acids to release elemental sulfur.

$$CO_2 + 2H_2 + H_2S \longrightarrow (CH_2O) + H_2O + 2S$$

Sulfate-reducing bacterial genera in soil include *Desulfomonas, Desulfovibrio*, and *Desulfatomaculum* (all obligate anaerobes). The most common and widespread species of these bacteria are *Desulfovibrio desulfuricans*, a non-spore-forming, obligate anaerobe which reduces sulfates at a rapid rate in waterlogged or flooded soils. In dry land soils, species of *Desulfatomaculum* which is a spore forming, thermophilic obligate anaerobes are responsible for reducing sulphates. An enzyme known as a "desulfurase" or a "bisulfate reductase" is excreted by all sulfate-reducing bacteria. Increased temperature, high organic matter content and increasing water levels (flooding) promote rate of sulfate reduction in nature.

Enzyme reactions in soil involving sulfur compounds

A. Sulphatases

Aryl and alkylsulfatase enzymes are thought to be crucial in sulfur mineralization since most of the soil organic sulfur is present as sulfate esters. The reaction can be expressed as

$R - OSO_3^- + H_2O$ --------> $R - OH + H^+ + SO_4^{2-}$

Based on the type of substrate, sulfatases are classified and mainly grouped into arylsulfatases, alkyl-sulphatases, mycosulphatases, glucosulphatases etc. Among all sulphatase enzyme Arylsulfatases or phenol sulfatases is most ubiquitous in soils of forest, cultivable land, marshes, sediment, etc. In soil, bacteria and fungi are the major source of this enzyme.

Methionine and cysteine, two of the major sulfur amino acids, are also transformed by enzymes in soil. It is important to note that first, cysteine oxidation to cystine (the disulfide form of the amino acid) occurs rapidly in soil since this reaction is catalyzed by trace amounts of a variety of metal ions. A disulfide called thiocysteine is formed when the enzyme cystathionine lyase acts upon cystine. The Thiocysteine then can react with a free sulfhydryl group to produce hydrogen sulfide (H_2S).

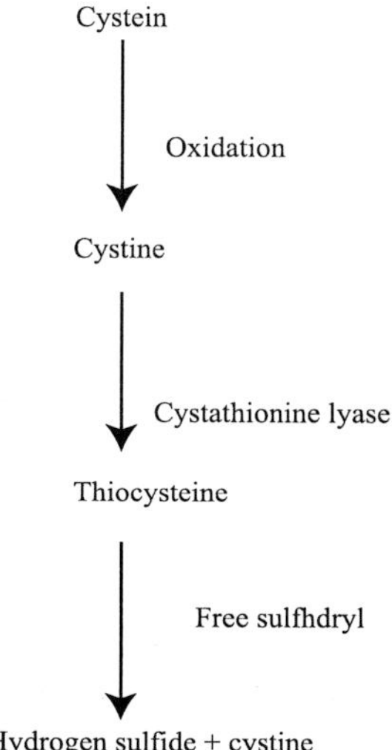

Observation of the chemical structures of cysteine and thiocysteine reveals how hydrogen sulfide may form during the reaction sequence described.

$HSSCH_2CH(NH_2)COOH + HSCH_2CH(NH_2)COOH$ ---> $H_2S + (SCH_2CH(NH_2)COOH)_2$

(thiocysteine) (cysteine) (hydrogen sulfide) (cystine)

Both cysteine and cystine (contains a free sulphydryl) may be present in an environments that are neither highly aerobic nor anaerobic. Experiments in the field have demonstrated that hydrogen sulfide losses are more likely to occur during the early stages of waterlogging than later after the development of a strong anaerobic environment, i.e. when sulfate reduction becomes the dominant mechanism of hydrogen sulfide production. An explanation for this reaction may be found in the series of reactions just described. According to the soil depth, soil type, organic matter content, season and climate, the activity of the enzyme arylsulfatase varies and its maximum activity is found in surface soils with pH ranged between 5.5 to 6.2. Various factors that influence microbial biomass and activity are also affecting the activity of the enzyme. The activity of arylsulphatase in soil correlates significantly with moisture percentage, clay content, rhizosphere type, organic carbon, nitrogen, and type of vegetation. The repeated application of S^0 fertilizer to soil has been observed to decrease enzyme activity due to a decrease in microbial populations and the inhibitory effect of large amounts of SO_4^{2-}.

B. Rhodanese

Another enzyme called rhodanese (thiosulfate cyanide sulfotransferase), belongs to the class transferase and involved in the sulfur cycle has been detected and characterized in soil. It catalyzes the following reaction between thiosulfate and cyanide in which thiocyanate is formed.

$S_2O_3^{2-} + CN^-$ ---------> $SCN^- + SO_3^{2-}$

In animal, plant tissue, bacteria and soils the enzyme is found. Significant rhodanese activity is observed in different soils. During the oxidation of elemental sulfur to sulphate, as intermediates both thiosulfate and tetrathionate are produced which are metabolized further under rhodanese-catalyzed reaction.

Groups of microorganisms involved in sulfur transformation

Sulfur is released by microbes from sulfide minerals and elemental sulfur at the earth's surface to soil. Sulfur becomes readily available in soil and for plants only from sulfate minerals, because plants can only take up sulfur in the form of $SO_4^{=}$. Sulfide S, elemental S, and thiosulfate S must be oxidized first. Although microorganisms from bacteria, eukaryotes (fungi) and archaea are involved in the sulfur oxidation; the bacteria, *Thiobacillus* sp. played the major role. Aerobic sulfur oxidation is restricted to *Sulfolobales*, an important member in the domain Archaea. Among the fungi *Aureobasidium pullulans, Alternari atenui* and *Epicoccum nigrum* and a range of *Penicillium* sp., *Myrothesium circutum, Selecobasidium constrictum,* and *Aspergillus* sp. are found to be associated in the oxidation of elemental-S and thiosulfate. Sulfur oxidizing bacteria can be categorized into three groups as follows:

a. Chemolithoautotrophs

For the growth and development of these bacteria, it uses the oxidation of sulfur and carbon dioxide as a source of energy and carbon, respectively. Examples of these group of bacteria are *Thiobacillus thioparus, T. denitificans, T. neapolitanus,T. ferrooxidans, T. halophilus* , *T. thiooxidans,* and some species of *Thiomicrospira.*

b. Chemolithoheterotrophs

These group of bacteria utilize oxidation of sulfur and organic molecules to obtain energy and carbon respectively, for its growth and development. The examples of chemolithoheterotrophs group of bacteria are, *Thiobacillus acidophilus, T. aquaesulis, T. novellus Paracoccus versutus, P.dentrificans* , *Thiospaera pantotroph, Xanthobacter tagetidis,* and *Thiomicrospira thasirae.*

c. Chemolithomesoptrophs

For growth and development, these bacteria utilize oxidation of sulphur and inorganic as well as organic molecules as a source of energy and carbon, respectively. Examples are *Thiobacillus denitrificans* and *T. ferrooxidans*. In sulfur oxidation, a number of enzymes are involved viz. thiosulfate dehydrogenase, trithionate hydrolase ,tetrathionate hydrolase and sulfur oxygenase. Whereas, under anaerobic conditions such as environment in low-land rice paddies, S reducing organisms reduce sulphates that produce H_2S, responsible for the bad odour emerged from paddy fields. To obtain energy, sulfate-reducing bacteria reduce sulfate. A total of 220 species of sulfate-reducing bacteria belong to 60 genera are known. The largest group (around 23 genera) consists Desulfovibrionales, Desulfpbacteriales and Syntrophobacterales. *Desulfotomaculum, Desulfosporosium and Desulfosporomusa* make up the second largest group.

Both aerobic and anaerobic bacteria can oxidize organic sulfur compounds. Their morphology can range from filamentous (*Beggiatoa, Thiothrix,* and *Thioploca*) to non-filamentous (*Thiobacillus*). There have also been reports of several fungi and actinomycetes oxidizing sulfur (*Aspergillus, Penicillium, Microsporeum*). The microorganism *Thiobacillus* deserves special attention since it produces sulfuric acid when elemental sulfur is added to soil resulting reduction in the soil pH as low as 2.0 after prolonged incubation with the bacterium. *Thiobacillus* may play a role in controlling plant diseases in sulfur amended soils in regard to potato scab caused by *Streptomyces scabies* and sweet potato rot caused by *S. ipomoea* has been demonstrated. In acidic soil (below pH 5.0), inoculation of soil with thiobacilli after addition of sulfur effectively reduces losses caused by these pathogens. Sulfur application with thiobacilli inoculation can also make alkali soils suitable for crop cultivation. The formation of H_2SO_4 in soil following additions of elemental sulfur augments nutrient mobilization by increasing the level of soluble phosphate, potassium, calcium, manganese, aluminium and magnesium. Adding sulfur to soil results in the formation of H_2SO_4 which increases nutrient mobilization by enhancing the level of soluble phosphate, potassium, calcium, magnesium manganese, and aluminium in the soil. Sulfur application can correct soil manganese deficiencies. By reducing inorganic sulfate into hydrogen sulfide, sulfate-reducing bacteria reduce plant availability of sulfur and thus affect crop yield. The obligate anaerobe *Desulfovibrio desulfuricans* belongs to this class of bacteria that produces hydrogen sulphide at a rapid rate. Inorganic sulfate reduction is also carried out by other species of Desulfovibrio, although the exat pathway is still unclear.

Role of mycorrhizal association in plant sulfur supply

Several fungi in soil can mineralize S from sulfate esters. By contrast, sulfonates, the predominant organo S source in soil, is mobilized by an exclusively bacterial multicomponent monooxygenase enzyme complex. It is likely that soil S cycling involves complex interactions between free-living and symbiotic microbial populations associated with roots. In The arbuscular mycorrhizal (AM) fungi form symbiosis with 80% of plant species which rely upon them for growth. In AM symbiosis, fungi penetrate root cortical cells, forming microscopic branched structures called arbuscules that increase the efficiency of metabolite exchange between plant and fungi. An extraradicular AM hypha provides a surface that is colonized by functional bacteria. The interactions between AM fungi and phosphorus (P) and nitrogen (N) mobilizing bacteria has been reported in a number of studies. Similar to S, plants rely on interactions with mycorrhizal fungi and associated microbes for mobilization of N and phosphorous. Research in recent years indicates that similar to the rhizosphere, considered as hot spot for microbial activity, mycorrhizosphere was also host a variety of bacteria that attach to fungal hyphae in order to mobilize organosulfur. Althogh, current findngs is not relecting any sulfatase and sulfonatase activity in arbuscular mycorrhiza, however their effect on the expression of plant host sulfate transporters is mentioned.

Objective type questions

A. Fill up the blanks with suitable words

i. Sulphur containing amino acids are ...., ... and ....

ii. The organic sulfur compounds thus can mainly be divided in two group namely .... and ....

iii. .... form of S may present in soil more than 95% of the total amount of S in temperate condition

iv. In arid or semiarid region, appreciable amount of ....form of sulfur present in soil.

v. ....the first stable compound of photosynthetic electron chain is a Fe-S protein

vi. The optimum N:S ratio for legume is ....

vii. In mature leaf tissue of healthy plants, S ranges from .... based on dry weight

viii. In the 1970s, ... were the first to report arylsulfatases.

ix. Organisms responsible for oxidation of elemental sulfur to sulphates are the members of genus ....

x. In bottom of water bodies containing sulfur material, ... group of bacteria are more commonly found

B. Multiple choice questions

i. Sulphate are reduced to hydrogen sulphide by

a. Desulfotomaculam sp. b. Thiobacillus thiooxidance

c. Photosynthetic sulphur bacteria d. Rhodospirrillum

ii. Purple and green sulphur bacteria use ....as the electron donor to reduce carbon dioxide

a. S^{2-} b. SO_4^{2-}

c. H_2S d. None of this

iii. Which is not a member of the genus green and purple bacteria

a. Chlorbium b. Rhodopseudomonas

c. *Desulfovibrio* d. *Chromatium*

iv. Colourless filamentous sulfur bacteria of the species, generally found in sulfide-containing waters and able to oxidize sulfide to sulphate are

a. *Thiothrix* b. *Beggiatoa*

c. *Thioploca* d. *All of this*

v. Sulfate present in the soil is assimilated and incorporated into proteins by plants and microorganisms,known as

a) assimilatory sulfate reduction

b) dissimilatory sulfate reduction

c) Sulphur mineralization

d) None of this

vi. Calcium sulfate, is attacked by members of the genus *Desulfovibrio* and *Desulfatomaculum* under anaerobic conditions to release

a) SO_2 b) H_2S

c) S d) None of this

vii.An enzyme is excreted by all sulfate-reducing bacteria known as

a. Desulfurase b. Bisulphate oxidase

c. Sulfurase d. None of this

viii.The C:S ratio in carbonaceous materials above which immobilization dominates mineralization is approximately

a. 20:1 b. 50:1

c. 100:1 d. None of this

ix. Thiosulfate and tetrathionate are metabolized further under the .... catalyzed reaction

a. Sulphurase b. Thiosulphate oxidase

c. Rhodanese d. None of this

x. The microorganism .... produces sulfuric acid when elemental sulfur is added to soil resulting reduction in the soil pH after prolonged incubation with the bacterium

a. Desulfovibrio b. Thiobacillus

c. Pseudomonas d. None of this

Answer keys

A. Fill in the blanks

i. cysteine, cystine, methionine **ii.** Ester sulphate group and carbon bonded sulphur group **iii.** organic **iv.** Inorganic **v.** Ferrodoxin, **vi.** 15-16:1 **vii.** 0.1-0.4% **viii.** Tabatabai and Bremner **ix.** *Thiobacillus* **x.** Green and purple

B. Multiple choice questions

i. a, ii. c, iii. c, iv. d, v. a, vi. b, vii. a, viii. b, ix. c, x. b

C. Descriptive type questions

i. What are the source of sulphur in soil.Write in brief S cycle in soil.

ii. Summerize sulphur transformation in soil.

iii. Write down the role of microbs in oxidation of inorganic sulphur.

iv. Write in brief role of sulphatases and Rhodanese enzyme in soil.

v. What are the difference between Chemolithoautotrophs and Chemolithoheterotrophs

15

Plant Growth Promoting Rhizobacteria and Use of Rhizobacteria as Biocontrol Agents: Biopesticides

Unlike root-free soil, the rhizosphere around growing plants is a very dynamic environment that hosts a much higher number of microbes, in particular bacteria. Microbes interact with each other and with plants in symbiotic, associative, neutral or antagonistic ways. A plant colonized or penetrated by a microorganism can either result in asymptomatic or disease or form an association or symbiosis, depending on the cells' perception of each other and how the environment affects that interaction. Microbes that penetrate and colonize plants have developed elaborate strategies to subvert plant defenses.

Plant growth promoting rhizobacteria (PGPR) or plant health promoting rhizobacteria (PHPR) refers to the group of beneficial, root associative bacteria that promote plant growth. The *Fluorescent pseudomonad*s belongs to PGPR's major group along with other bacteria such as Acetobacter, Agrobacterium, Actinoplanes, Arthrobacter, Alcaligenes, Azotobacter, Azospirillum, Bacillus, Clostridium, Cellulomonas, Erwinia, Enterobacter, Flavobacterium, Pasteuria, Rhizobium, Serratia and Xanthomonas. Rhizobia and bradyrhizobia, which form a symbiotic relationship with leguminous plants, are also among the beneficial microorganisms in the rhizosphere. Plant growth may be impaired if the rhizosphere lacks an appropriate microbial population. Plants may get benefit from microbial populations in the rhizosphere in a variety of ways, including:

- increased recycling, solubilization and uptake of mineral nutrients.
- synthesis of amino acids, vitamins, and plant hormones viz. auxins and gibberellins which promote plant growth and
- development of antagonistic relationships with potential plant pathogens based on antibiotic production and competition. Thus, rhizobacteria, capable of protecting plant roots from being affected by pathogens, are considered as ideal biocontrol agents.

(I) Plant growth promotion

(i) Increased availability and uptake of mineral nutrients

In the rhizosphere, microorganisms contribute to plant nutrition by either (a) limiting concentrations of inorganic nutrients before they travel to plant roots and sometimes by (b) increasing availability of inorganic nutrients to plant roots. Many rhizosphere microorganisms possess the capability to mineralize the bound or fixed form of nutrients in the soil organic matter and causes decomposition of cellulose, hemicellulose and lignin; solubilize the bound phosphorus; causes oxidation-reduction or mineralization of sulphur compounds. Some rhizosphere bacteria carry out fixation of atmospheric inert nitrogen into ammonia using the enzyme nitrogenase.

Biological nitrogen-fixing systems

Rhizosphere bacteria contribute significantly to biological nitrogen fixation and thus beneficial for soil and plant health, besides nutrient mobilization, solubilization and plant growth promotion through secondary metabolites. On a global basis, among the nitrogen-fixing systems the legume-Rhizobium symbiosis alone accounts for 70-80% of total nitrogen fixed biologically and one third of total nitrogen input required by world agriculture each year. Despite having a limited potential for nitrogen input on an acreage basis, free living/ associative diazotrophs now appear in almost all ecological environments and have a greater impact on nutrient efficiency and crop physiology than previously thought. Similarly, cyanobacteria, the free-living nitrogen fixers in aquatic environment, specifically in rice fields, contribute about one-third of the nitrogen requirement of the crop.

Table on An estimate of the average rate of biological nitrogen fixation by diazotrophs and associations.

Sl.No.	Organism/System	N_2 fixed (kg ha^{-1} yr^{-1})
1.	Free living microorganisms - *Azotobacter, Clostridium, Derxia*	0.1-25
2.	Associative symbioses - Azospirillum, Azoarcus, Herbaspirillum	5-36
3.	Cyanobacteria - Anabaena, Nostoc, Oscillatoria	15-50
4.	Azolla-Anabaena symbiosis	313
5.	Rhizobium-legume symbiosis	57-600
6.	Nodulated non-legumes - *Casuarina-Frankia, Parasponia-Rhizobium*	2-300

(i) ***Phosphate solubilization and mobilization:*** Phosphate solubilizing and mobilizing microorganisms comprise bacteria (*Pseudomonas, Bacillus*, nitrogen-fixing cyanobacteria, other free-living diazotrophs and actinomycetes) as well as the fungi (ectomycorrhizae, *Aspergillus, Penicillium, Paecilomyces* etc.). Arbuscular mycorrhizal fungi facilitate phosphate nutrient transfer to the higher plants. The microorganisms in the rhizosphere increase phosphate availability by solubilizing materials that otherwise would not be available to the plant. Presence of these microorganisms in the rhizosphere facilitate the activity 10 times more than when they are present as free-living saprophytes because of easy availability of organic soluble compounds in root exudates for their growth. The uptake of phosphorus by plants increases when they are associated with rhizosphere microorganisms than in sterile soils. Acids produced by microbial activity that dissolve apatite result in the release of soluble forms of phosphorus and ultimately increase phosphate availability. By acidifying the soil's rhizosphere, microbially-produced organic acids can solubilize essential minerals. Study of 2-ketogluconic acid production by Gram-negative bacteria growing in glucose media illustrates the ability of organic acids excreted by soil bacteria to solubilize soil minerals. Moreover, acidification of the rhizosphere environmental conditions resulting from metabolically produced hydrogen ions changes the pH sufficiently so that soil minerals can be mobilized.

Insoluble salts and minerals, either natural or synthetic have been shown to dissolve a variety of metals that it contains. Various bacteria, including common soil organisms, synthesize siderophores and other metal chelators. Rhizosphere microorganisms that produce chelating agents (e.g. production of siderophore) increase the solubility of iron and manganese compounds and thus iron and manganese may be more available to plants.

It has also been shown that microorganisms on roots significantly increase the uptake rates of calcium by the roots. This increase may be due to the high concentrations of CO2 in the rhizosphere produced by microorganisms, which increases the solubility, and thus the availability of calcium. Translocation of various radiolabelled organic compounds and heavy metals along mycelial filaments has also been demonstrated.

(ii) ***Synthesis of vitamins, amino acids, auxins and gibberellins:*** Auxins, cytokinins, and gibberellin-like compounds are synthesized by microorganisms and these compounds promote seed germination rate and stimulate root tissue development, thus increasing the root system's ability to supply nutrients and water for production of above ground biomass.For example, the wheat seedlings rhizosphere are habitat of a significant quantities of bacteria that able to produce a plant growth hormone namely indole acetic acid (IAA) that increase the growth of plant roots. In older wheat plants, there are a lower proportion of microorganisms in the rhizosphere capable of producing IAA. This may be in response to a decline in root exudates production, but it also beneficially corresponds to a decreased need for the growth hormone by the plant.

(iii) ***Development of a stable soil structure:*** Indirectly, but nonetheless clearly beneficial, results of root invasion that stimulate soil community development are creation of stable soil structures that promote plant community development by stimulating microbial communities in the rhizosphere. Polysaccharide material, such as capsules and slimes, is produced in the rhizosphere community to bind soil minerals into microaggregates. In addition, soil structure is improved through fungal mycelial production and by soil particles' association with root tissue. Increased soil aggregation results improved soil structure that favours positively in soil aeration, water infiltration, and root penetration.

(iv) ***Release of allelopathic substances:*** Microorganisms in the rhizosphere release allelopathic (antagonistic) substances that allow plants to establish amensal (antagonistic) relationships with other plants. Some allelopathic substances surrounding some plants may responsible for prevening the habitat to be established by other plants, and this may represent a synergistic relationship between a plant and its rhizospheric microbial community. Rhizosphere bacterial populations of young wheat plants have been shown to inhibit the growth of pea and lettuce plants. As wheat plant matures, the proportion of these bacteria decreases and they are replaced by a higher proportion of microorganisms capable of producing growth promoting substances similar to gibberellic acid. In the same way, rhizosphere bacteria (rhizobacteria) can suppress growth of sensitive plants. The *Pseudomonas sp.*, for instance, has been isolated from winter wheat (*Triticul aestivum* L.) and downy brome (*Bromus tectorum* L.) roots, and it is able to inhibit growth of downy brome weeds.

(v) ***Rhizobacteria as biocontrol agents:*** A method of controlling soil-borne plant pathogens using microorganisms, natural or modified, genes or any gene products to control the adverse effect by any undesirable organisms or pests, and to promote desirable organisms, viz. trees, crops, beneficial insects and microorganisms and animals, is known as biocontrol. As rhizosphere bacteria are capable of providing the front line defense to plant roots against the attack of pathogens, they are ideal for use as biocontrol agents. A biocontrol agent suppresses disease through interactions among the biocontrol agents and the members of the rhizosphere, spermosphere, or phyllosphere community.

Invasion of the plant tissue by the pathogen was directly or indirectly inhibited by the rhizosphere microflora. . Disease reduction may be due to competition between pathogen and root microbes, antagonism between root microbes and pathogen, or alteration of root exudate diffusion into soil, which may interfere with chemotaxis of pathogen to the root.

(a) Competion between root-inhabiting microbes and plant pathogens may occur for nutrients, space, or for binding sites on the root surface.. Competition for space and nutrients may impair a pathogen's ability to establish critical population densities for disease initiation, while competition for specific binding sites may reduce a pathogen's ability to begin infection process.

(b) The antibiotics and siderophores produced by rhizosphere microbs soil may confer survival advantage. The production of antibiotics and siderophores by various *Pseudomonas* strains have been shown to reduce the incidence of take-all disease in wheat roots.

(c) Furthermore, rhizosphere microbes act to interfere with the pathogen's chemotactic attraction to root receptor sites and thus reducing the incidence of root disease incidence. The importance of chemotactic interactions in the production of plant diseases is harder to document in the field, but they are easily deduced. Plant pathogens may be attracted to root exudates that contain a number of compounds.By growing root inhabitants (including mycorrhizal fungi), the quantity and diversity of organic compounds diffused from the roots are reduced, thereby decreasing the likelihood of encountering a pathogen.

Rhizobacteria's major disadvantages as biocontrol agents are their variable performance in the field and the need for precautions to ensure their survival and delivery.Additionally, a biocontrol agent's effectiveness may be limited to a specific location due to effects of soil and climate.Furthermore, biological control is dependent on an established and maintained of threshold population of bacteria on planting material or in soil, and when viability drops below the threshold level, the possibility of biological control disappears.. A number of soil edaphic factors, including soil moisture, temperature, clay content, pH, interactions between the disease control microorganisms and other rhizosphere bacteria and with pathogens, will also influence their ability to survive and to tolerate adverse conditions.

(II) Use of Plant growth promoting rhizobacteria as biocontrol agents : Biopesticides

A number of rhizosphere bacteria are capable of controlling various root, foliage and post-harvest diseases of agricultural crops. These rhizobacteria are an ideal biocontrol agent since they can provide the plant roots with an early line of defense against various plant pathogens. The objective of biocontrol is to improve plant health by harnessing disease-suppressing microorganisms.. The interaction between a plant, a pathogen, a biocontrol agent, a microbial community on and around the plant, and the physical environment leads to the suppression of disease. A few non-pathogenic rhizobacteria may also cause physiological changes that make plants more resistant to pathogens. There are various advantages to biological disease control organisms, namely: (1) It is considered safer to use these organisms than many of the chemicals currently used, (2) Non accumulation in food chain (3) repeated applications can be avoided due to self replication ability, (4) Minimal chances for target organism to develop resistance as contrast to the application of chemical control agents, (5) Can be integrated with a chemical control agent in case of less efficiency than, and (6) Biocontrol agents that have been developed properly aren't considered harmful to the ecology.

Suppression of growth of pathogenic microorganisms

Rhizobacteria have been found to suppress the disease caused by various pathogenic bacteria and fungi and have the potential for use as biocontrol agents. Biological control can be defined as the control of a plant disease with a natural biological process. The rhizosphere bacteria are ideal for use as biocontrol agent as they can provide the front line defense for plant roots against the attack by various plant pathogens. Disease suppression by biocontrol agents occurs due to interactions among the biocontrol agents with members of the rhizosphere or phyllosphere community.

The mechanisms by which rhizobacteria inhibit the growth of phytopathogenic microorganisms are not fully understood, but are thought to include: (i) antibiotic production, (ii) production of siderophores, (iii) production of hydrolytic enzymes such as β-1,3-glucanase, chitinase and cyanide, and (iv) phytoalexins production and induction of systemic resistance. One or more antibiotics have been shown to play role in disease suppression in many biocontrol systems by mutant analyses and biochemical studies using purified antibiotics. Antibiotics may be simply a highly effective mechanism for suppressing pathogens in the rhizosphere.

Table on Biocontrol of pathogens by antagonistic microorganisms

Biocontrol bacteria	Pathogen suppressed	Crop
Pseudomonas fluorescens	*Erwinia* sp.	Potato
	Fusarium sp.	Radish
	Thievaloviopsis basicola	Tobacco
	Rhizoctonia solani	Peanut
	Fusarium oxysporum sp. *Ciceri, Macrophomina phaseolin*a	Chickpea
	Pythium ultimum	Pea
	Xanthomonas malvacearum	Cotton
	Gaeummanomyces graminis var. tritici	Wheat
P. putida	*Fusarium* sp.	Radish
	E. carotovora, X. campestris	Potato
	Fusarium oxysporum	Flax
	Fusarium oxysporum f. sp. *lycopersici*	Tomato
P. aureofaciens	*Gaeummanomyces graminis var. tritici*	Wheat
	Botrytis cinerea, P. expansum	Apple
	Heterodera glycines, Meloidogyne incognita	Soybean
Pseudomonas sp.	*F. moniliforme*	Maize
	Pythium ultimum	Sugarbeet
	Rhizoctonia solani	Cowpea
	Agrobacterium tumefaciens	Grapevine
Bacillus cereus	*Alternaria solani, Pythium*	Potato
B. polymixa, B. pumilis	*G. graminis var. tritici*	Wheat
Bacillus subtilis	*Fusarium roseum*	Corn
Bacillus sp.	*G. graminis var. tritici, Rhizoctonia*	Wheat
Others		
Burkholderia cepacia	*Aphanomyces, Botrytis, Fusarium, Monilina, Penicillium, Pyricularia, Ralstonia*	
Agrobacterium radiobacter	*A. tumefaciens*	
Azotobacter chroococcum	*Rhizoctonia solani*	
Rhizobium/Bradyrhizobium sp.	*Macrophomina phaseolina, Rhizoctonia solani, Fusarium*	
Enterobacter cloacae	*Alternaria, Fusarium, Pythium, Rhizoctonia, Sclerotinia*	
Erwinia sp.	*Fusarium, Sclerotinia*	

(i) ***Production of antibiotics:*** Antibiotic production by rhizobacteria is one of the major mechanisms postulated for antifungal activity and plant growth promotion. These antimicrobial compounds may act on plant pathogenic fungi by inducing fungistasis, inhibition of spore germination, lysis of fungal mycelia, or by exerting fungicidal effects. A large number of antibiotics including diacetylphloroglucinol, oomycin A, phenazines, pyocyanine, pyrroles, pyoluteorin and pyrrolnitrin etc. are produced by rhizobacteria which help in suppression of pathogen growth.

The antibiotics clearly implicated in biocontrol by fluorescent pseudomonads were the phenazine derivatives that contribute to disease suppression by *Pseudomonas fluorescens* strain 2-79 and P. aureofaciens strain 30-84, which control take-all of wheat. Evidence for the role of phenazines include an analysis of transposon insertion mutants that lack the ability to produce phenazine-1-carboxylate are reduced in disease suppressiveness. P. fluorescens strain CHA0 isolated from tobacco rhizosphere was found to produce a variety of secondary metabolites i.e. pyoluteorin and 2,4-diacetylphloroglucinol, hydrogen cyanide, salicylic acid, pyochelin and pyoverdine and protected various plants from diseases caused by soil borne pathogenic fungi.

Trichoderma and Gliocladium are closely related fungal biocontrol agents. Each produces antimicrobial compounds and suppresses disease by diverse mechanisms, including the production of the structurally complex antibiotics gliovirin and gliotoxin. Mutants of Gliocladium virens that do not produce gliotoxin are reduced in their ability to control *Pythium* damping-off. Mutants with increase or decreased antibiotic production show a corresponding effect on biocontrol.

(ii) ***Production of siderophores:*** Iron is essential element for all living organisms, with the possible exceptions of certain lactobacilli. Iron is abundant in Earth's crust, but most of it is found in the highly insoluble form of ferric hydroxide; thus, iron is only available to organisms at concentrations at or below 10^{-18} M in soil solutions at neutral pH. This presents a challenge for bacteria, which require iron at micromolar concentrations for growth. To cope with its solubility, many microorganisms synthesize extracellular siderophores in response to low iron stress. Siderophores are low molecular weight, high affinity iron chelators that transport iron into bacterial cells. Nearly all aerobic and facultative anaerobic bacteria have been found to produce siderophores. PGPR produce different types of siderophores, which are involved in disease suppression and plant growth promotion. Two important siderophores-mediated iron uptake systems have been found in the rhizobacteria: one involving the fluorescent pseudobactin (pyoverdine) and other pyochelin. Plant growth promoting rhizobacteria (PGPR) appear to exert their beneficial effects in part by producing extracellular siderophores under iron limiting conditions that efficiently chelate environmental iron, making it less available to endemic microorganisms, thus inhibiting their growth. The various categories of siderophores produced by PGPR include catechol, hydroxamate, pyoverdine and some other types like azotochelin, anthranilic acid and azotobactin. Siderophore-negative mutants were derived from P. fluorescens strain 3551 and B224 by chemical, Tn5 insertion and UV mutagenesis. Sid^- mutants of strain 3551 provided less biocontrol than parent strain against Pythium ultimum causing damping-off disease of cotton, whereas Sid^- mutants of strain B224 showed less increase in growth of wheat than Sid^+ parent strain in presence of wheat pathogen P. ultimum var. sporangiferum.

An interesting aspect of siderophore biology is that diverse organisms can use the same type of siderophore. Microorganisms may use each other's siderophores if they contain the appropriate uptake protein and plants can even acquire iron from certain pseudobactins.

(iii) ***Production of other secondary metabolites and enzymes:*** Besides antibiotics and siderophores, a number of other metabolites are also produced by PGPR, which play important roles in plant growth promotion and resistance to diseases in plants. Lysis by hydrolytic enzymes excreted by microorganisms is a well-known feature of mycoparasitism. The process of destruction of pathogens by the action of cell wall lysing enzymes is known as parasitism. Extracellular chitinase and laminarinase produced by *P. stutzeri* have shown marked effect on mycelial growth inhibition rather than spore germination and also caused lysis of *F. solani* mycelia and germ tube. *P. cepacia* decreased the incidence of disease caused by Rhizoctonia solani, Sclerotium rolfsii and Pythium ultimum due to production of β-1,3-glucanase (laminarinase).

Hydrogen cyanide (HCN) is the other secondary metabolite, which is known to be produced by many rhizosphere bacteria and has been demonstrated to play a role in the biological control of the pathogens. HCN overproducing bacterial strains resulted in small but statistically significant increase in the suppression of symptoms caused by

Mycophaerella graminicola and *Puccinia recondita* f. sp. *tritici* on wheat seedling leaves. Volatile compounds such as ethylene and ammonia gas present in the soil atmosphere have also been reported to inhibit germination of fungal spores. But, the production of volatile inhibitors by disease-control bacteria is probably not a significant mechanism in the control of root pathogens.

(iv) ***Phytoalexins production and induction of systemic resistance:*** Some non-pathogenic rhizobacteria can induce physiological changes in the plants due to production of phytoalexins, making them more resistant to pathogens. Phytoalexins are low molecular weight antimicrobial compounds that are synthesised by and accumulated in plants after exposure to microorganisms. Most phytoalexins are flavonoids or isoflavonoid-like compounds and their occurrence is widespread in the plant kingdom. Phytoalexins synthesis can be used as indicators of enhanced defense mechanism in bacteria-treated plants. Some polyphenolic compounds have been identified in root exudates of legumes grown under sterile conditions. In lentil, three desoxy-5 flavones and in soybean two isoflavonoids, coumastrol and diadzein were found. Higher phytoalexins concentration was observed in nodules of field grown plants than that from sterile pot cultured plants inoculated with different R. leguminosarum strains, suggesting that a plant defense mechanism is induced under field conditions. Flavonoid molecules secreted by legume roots act as inducers for the transcription of rhizobial nodulation genes and thus help in enhancement of nodulation. In beans, coumestrol is released in response to both pathogens and symbiotic rhizobia, in the later case, it induces the nodulation genes at roughly the same concentration as do other nod-gene inducers. Some biocontrol ganets induce a sustained change in the plant, increasing its tolerance to infection by a pathogen, a phenomenon known as induced systemic resistance (ISR). In some cases, it is clear that induced resistance by biocontrol agents involves the same suite of genes and gene products involved in the well-documented plant response known as systemic acquired resistance (SAR), but this is not always the case. SAR is typically a response to a localized infection or an attenuated pathogen, which is manifested in subsequent resistance to a broad range of other pathogens.

Various non-pathogenic rhizobacteria have the ability to induce a state of systemic resistance in plants, which provides protection against a broad spectrum of phytopathogenic organisms including fungi, bacteria and viruses. Induced resistance brought about by prior inoculation of the host by a pathogen, avirulent or incompatible forms of a pathogen, or heat killed pathogens has been attributed to induced physiological response of the host plant against subsequent inoculation by the virulent pathogens. Induced systemic resistance in plant has been demonstrated in over 25 crops, including cereals, cucurbits, legumes, solanaceous plants and trees against a wide spectrum of pathogens. Inoculation of PGPR strains 98B-27 (*P. putida*) and 90-166 (*Serratia marcescens*) and the pathogen (*F. oxysporum* f. sp. *cucumerinum*) on separate halves of roots of cucumber seedlings exhibited that both PGPR strains induced systemic resistance against Fusarium wilt ass expressed by delayed disease symptoms and reduced number of dead plants compared to the non-bacterized *F. oxysporum* f. sp. *cucumerinum*-inoculated plants. It has been shown that the biocontrol agent *P. fluorescens* strain CHAO induces SAR-associated proteins, confers systemic resistance to a viral pathogen, and induces accumulation of salicylic acid, which plays a role in signal transduction in SAR.

Biotic and abiotic factors affecting rhizosphere colonization

Several biotic and abiotic factors affect rhizosphere colonization by the rhizobacteria. These factors include the microbial and plant characteristics as well as the soil and environmental factors. The bacteria in the root zone are influenced by the root exudates (organic compounds), which act as nutrients as well as chemoattractants.

Biotic and abiotic factors and rhizosphere colonization.

1. Microbial characteristics

- Chemotaxis towards seed or root exudates
- Growth rate - bacteria *vs* fungi
- Antibiotic production - provide dominance over pathogens
- Siderophore production - chelation create iron deficiency for pathogens
- Cell surface properties like OMP, EPS, appendages
- Tolerance to fungicides or other chemicals
- Tolerance to dehydration

2. Plant characteristics

Plant species- root surface area

Variation in plant genotypes

3. Environmental and soil factors

- Nutrient availability
- Organic nutrients favour oligotrophs
- Inorganic nutrient sufficiency favours pathogens
- Soil type, texture and structure, moisture, pH and temperature
- Soil atmosphere,
- Applied pesticides

In the colonization process, the bacteria with faster growth rate and oligotrophic nature colonize the root surface. Plant characteristics, like root structure and plant genotypes also influence the colonization. Environmental factors like available nutrients, soil properties, application of pesticides etc. affect rhizosphere colonization by the beneficial rhizobacteria. Normally organic environment in soil favours the beneficial flora while inorganic nutrient sufficiency favours the pathogens..

Beneficial effects of rhizobacteria in legumes

Legumes are widely used for food, fodder, fuel, shade, timber, green manure and as cover crops in different agricultural systems. These crops meet their nitrogen requirement through symbiotic nitrogen fixation by forming root nodules with rhizobia. Legume rhizosphere provides most of the nutritional requirements of nodule bacteria and enhances the population several folds, which becomes maximum during the plant growth season. Besides rhizobia, a large number of other rhizosphere bacteria have also been identified in the legume rhizosphere among these *Pseudomonas* and *Bacillus* sp. are predominantly found in rhizosphere of cultivated legumes.

Co-inoculation of rhizobia with such rhizobacteria in legumes has been found to enhance nodulation, increase in root length and gain in plant biomass and yield. Such effects have been observed by co-inoculation of *Pseudomonas putida* with *R. phaseoli* in *Phaseolus vulgaris*, *R. japonicum* with *P. fluorescens*, *Bacillus cereus,* or Serratia liquefaciens in soybean, R. meliloti with P. syringae in alfalfa, R. leguminosarum with a toxin releasing Pseudomonas or P. fluorescens in pea, R. trifolii with Pseudomonas in clovers, Rhizobium with

Bacillus subtilis in pigeon pea, Rhizobium with Pseudomonas and Bacillus in chickpea.

Beneficial effects of PGPR in non-legume crops

Microbial activity in the plant rhizosphere has substantial effects on growth and yield of non-leguminous crops. A large number of rhizobacteria have potential for use as inoculants in cereal crops and the plants inoculated with these bacteria have shown increased grain yield and plant biomass accumulation. The beneficial effects of these bacteria are attributed to increased nitrogen input through biological nitrogen fixation, production of plant growth hormones, increased root length and volume, enhancement in nutrient uptake, increased nitrate reductase activity in plants, production of antibacterial and antifungal compounds and reduced insect and disease infestation. The most common PGPR for cereals, vegetables and oilseeds include *Bacillus* sp., *Bacillus subtilis, Pseudomonas fluorescens, P. cepacia* and *P. putida* group. These groups are also involved in suppression of root borne diseases and have been used as biocontrol agents for plant diseases in onion, groundnut, chickpea, soybean and tomato.

Besides Pseudomonas and Bacillus species, other bacteria having N_2-fixing activity have been found in the rhizosphere of non-leguminous plants. The free-living diazotrophs were the first to be recognized and in the hope of increased crop yields with N_2-fixing bacteria, millions of hectares bearing various crops were treated with *Azotobacter sp.* Much of the early work with inoculation was carried out in Russia, Eastern Europe and India. Only about one third of field trials showed any positive effects. Increase in yields with Azotobacter inoculation have been observed from 2 to 45% in vegetables, 7 to 28% in cotton and 9 to 24% in sugarcane.

The genus *Azospirillum* is known to penetrate the fibrous root of plants and to grow intercellularly to some extent as well as, growing in the rhizosphere. Experimental work has shown that inoculation of wheat with A. brasiliense and A. lipoferum can increase significantly the yields of foliage and grain. In inoculation trial with pearl millet in India, average increase in grain yields with inoculation was higher by 11% in case of A. lipoferum and 8% in case of Azotobacter chroococcum over the uninoculated controls.

Approaches to increase the efficiency of PGPR

Now it has been well established that root colonization by PGPR is the prerequisite to enhance plant growth. Root colonization by introduced bacteria could be improved by increasing the population size, distribution or survival of bacteria, along with manipulation of soil factors that may positively or negatively affect colonization, as well as bacterial traits that may contribute to rhizosphere competence. Cell surface characteristics influence attachment to roots, which may be necessary for colonization. Certain mutations that affect accumulation of secondary metabolites also influence colonization of plant material in field soil. A promising approach that will likely broaden the array of traits considered to be important for colonization is to screen mutants directly for increased or decreased ability to colonize roots. Mutants of *Pseudomonas* strains of both genotypes have been identified, and analysis of these mutants indicated that phototrophy for amino acids and vitamins, rapid growth rate, utilization of organic acids and lipopolysaccharide properties contribute to colonization ability. On the basis of identified colonization genes and traits, colonization mutants of P. chlororaphis strain PCL1391 have been constructed. These mutants have lost their ability to control tomato root rot showing that root colonization is an essential trait for biocontrol.

Recently, genes of *Pseudomonas* biocontrol strains have been identified that can be induced or repressed by the presence of phytopathogenic fungi. *in vivo* expression technology (IVET) has been used to show that the presence of *Phytophthora parasitica* can induce various genes in P. putida, including genes encoding for diacylglycerol kinase, ABC transporters and outer membrane porins. In contrast, two ribosomal RNA operons of P. fluorescens were found to be repressed by pythium ultimum. The biocontrol performance of soil pseudomonads may be improved by the introduction of antibiotic biosynthetic genes. Recombinant DNA strains with greatly increased DAPG and PCN production have been constructed. bacteria are being constructed which combine two of the three useful traits: production of DAPG and PCN; and ISR. The production of DAPG and PCN will be placed under the control of strong promoters or of exudate-induced or rhizosphere-induced promoters. Moreover, the genes responsible for the production of secondary metabolites and involved in plant growth promotion could be transferred to other rhizobacterial strains possessing good colonizing and competitive ability. A good PGPR strain should produce the secondary metabolites under variable growth conditions. Therefore, such strains should be selected, which show a constant and medium-independent production of secondary metabolites. Further, it has been found that PGPR show greater and more consistent disease suppression when applied as mixtures of ecologically diverse strains with similar functions.

Production of a high level of antifungal metabolites (AFMs) poses a physiological burden on the producing cell and results in a slower growth rate and, consequently result in a less competitive strain. Finally, combining biocontrol traits of several strains into one cell sometimes led to increased level of biocontrol. However, in a number of cases the results were negative, presumably because of metabolic interference of biosynthetic pathways. Using rhizosphere- or exudate-inducible promoters, the production of AFMs can be limited to the plant root, the only site where AFM is needed. This ensures minimal loss of energy, and therefore competitive force, from the producing microbe.

Constraints in the use of PGPR

The major disadvantages in using the PGPR as a biocontrol agent include variability of field performance and the necessity for precautions to ensure survival and delivery of the product. Also, the effectiveness of a given biocontrol agent may be restricted to aspecific location, due to the effects of soil and climate. Many soil edaphic factors, including temperature, soil moisture, pH, clay content, interactions of biological-disease control microorganisms with other rhizosphere bacteria and with pathogens will also affect their viability and tolerance to adverse conditions once applied. During root colonization by introduced bacteria, introduced microorganisms have to compete with indigenous microflora for carbon source, mineral nutrients and infection sites on the roots. Sometimes, this competition is so severe that introduced microorganisms fails to survive in the soil. Another factor that can contribute to inconsistence performance of PGPR is variable production or inactivation *in situ* of bacterial metabolites responsible for plant growth promotion.

Perspectives

The complex interactions between the PGPR, the pathogen, the plant and the environment are responsible for the variability observed in disease suppression and plant growth promotion. Various factors like nutrient uptake, production of vitamins and hormones, production of antibiotics, siderophores, HCN or other secondary metabolites are directly or indirectly involved in plant growth promotion. The inconsistency in performance of these PGPR strains is a major constraint to their wide spread use as biocontrol agent in commercial agriculture. However, genetic manipulation of PGPR has the potential to construct significantly better strains with improved biocontrol efficacy. Further the efficacy of biocontrol bacteria can be improved by developing the better cultural practices and delivery systems that favour their establishment in the rhizosphere. The applications of mixture of biocontrol agents may be a more ecologically sound approach because it may result in better colonization and better adaptation to the environmental changes occurring throughout the growing season. Future strategies are required to clone genes involved in the production of antibiotics, siderophores and other metabolites and to transfer these into the strains having the good colonization potential along with other beneficial characteristics such as nitrogen fixation. In near future, the biotechnological approaches used in manipulation of bacterial traits will lead to improved biocontrol activity and better plant growth.

From the perspective of developing nations, these are exciting strategies that may help to increase yields while avoiding some of the costs and environmental problems that come with the use of fertilizers, pesticides, herbicides and fungicides. However, most of the microbial biodiversity in soil remains unexplored and much work remains to be done to first identify and then characterize microorganisms that could be used in such applications. Furthermore, such approaches require a detailed knowledge of the molecular signalling that takes place between plants and microbes to drive expression of desirable traits and suppress unwanted effects in a controlled manner. Exploiting plants and microbes by using such an integrated approach requires a system biology strategy to understand the degree and complexity of plant-microbe interactions through the application of modern genomics technologies. Applying knowledge about beneficial plant-microbe interactions in the rhizosphere to plant breeding and genetic engineering technologies may allow us to increase food production while reducing stress on the environment and on global biodiversity.

Objective type questions.

A. Fill up the blanks with suitable words

i. .... are the free-living nitrogen fixers in aquatic environment, specifically in rice fields, contribute about one-third of the nitrogen requirement of the crop.

ii. Rhizosphere bacteria carry out fixation of atmospheric inert nitrogen into ammonia using the enzyme ....

iii. Frankia fix N2 in association with the nodulated non leguminous tree ....

iv. .... type of mycorrhizal fungi facilitate phosphate nutrient transfer to the higher plants.

v. .... are low molecular weight, high affinity iron chelators that transport iron into bacterial cells.

vi. Rhizosphere are habitat of a significant quantities of bacteria that able to produce a plant growth hormone namely .... that increase the growth of plant roots.

vii. The production of antibiotics and siderophores by various .... strains have been shown to reduce the incidence of take-all disease in wheat roots.

viii. Biocontrol of Fusarium sp. in radish is possible by using the antagonistic microorganism ....

ix. ....and .... are closely related fungal biocontrol agents.

x. ....is a N2 fixing, gram-negative, microaerophilic, bateria that known to penetrate the fibrous root of plants and to grow intercellularly to some extent as well as, growing in the rhizosphere.

B. Multiple choice questions

i. Which of the following is incorrectly matched?

a. Alnus – *Frankia* b. Alfalfa – *Rhizobium*

c. Mycorrhiza – *Rhodospirrilum* d. Azolla-Anabaena

ii. Which of the following is not a biofertilizer?

a. Mycorrhiza b. *Rhizobium*

c. *Agrobacterium* d. *Azotobacter*

iii. Which of the following is commonly used as a nitrogen fixer in paddy fields?

a. *Frankia* b. Agrobacterium

c. *Azospirrilum* d. *Rhizobium*

iv. Which of the following is an endomycorrhiza?

a. *Rhizobium* b. *Agaricus*

c. *Glomus* d. *Nostoc*

v. Which of the following is a Phosphate solubilizing and mobilizing microorganisms

a) *Pseudomonas* b) *Bacillus*

c) *Aspergillus* d) all of this

vi. Biocontrol of Rhizoctonia solani by the microorganism

a) Azotobacter chroococcum
b) *Bacillus* sp.
c) Pseudomonas putida
d) Both a and b

vii.Roll of PGPR includes

a. Synthesis of vitamins, amino acids, auxins and gibberellins
b. Release of allelopathic substances
c. Phosphate solubilization and mobilization
d. All of this

viii.Which is a slow growing rhizobia

a. *Rhizobium. leguminosarum*
b. *Sinorhizobium meliloti*
c. *Bradyrhizobium japonicum*
d. All of this

ix.Bacterial strains, able to suppress symptoms caused by *Mycophaerella graminicola* and *Puccinia recondita* f. sp. *tritici* on wheat seedling leaves, that produce

a. HCN
b. IAA
c. Catechol
d. None of this

x. Which is an antibiotic produced by rhizobacteria

a. Diacetylphloroglucinol
b. Oomycin A
c. Pyrroles
d. All of this

Answer keys:

A. Fill in the blanks:

i. Cyanobacteria, **ii.** Nitrogenase **iii.** Casuarina **iv.** Arbuscular **v.** Siderophores **vi.** Indole acetic acid (IAA) **vii.** *Pseudomonas* **viii.** Pseudomonas fluorescens/ putidaix. Trichoderma and Gliocladium **x.** Azospirillum

B. Multiple choice questions

i. **c** ii. **c** iii. **c** iv. **c** v. **d** vi. **d** vii. **d** viii. **c** ix. **a** x. **d**

C. Descriptive type questions

i. What is Plant growth promoting rhizobacteria (PGPR)? How it benefited plant growth?

ii. How phosphate solubilization and mobilization in rhizosphere was influenced by Plant growth promoting rhizobacteria (PGPR)?

iii. What is biocontrol agents? Whether Rhizobacteria can act as biocontrol agents? justify

iv. Write in brief several biotic and abiotic factors affecting rhizosphere colonization by rhizobacteria.

v. How to increase the efficiency of PGPR?